Diverse Hydrogen Sources for Biomass-derivatives Conversion

Zhibao Huo

Diverse Hydrogen Sources for Biomass-derivatives Conversion

Reaction and Mechanism

Springer

Zhibao Huo
College of Marine Ecology
and Environment
Shanghai Ocean University
Shanghai, China

ISBN 978-981-99-1672-6 ISBN 978-981-99-1673-3 (eBook)
https://doi.org/10.1007/978-981-99-1673-3

This Springer imprint is published by the registered company Springer Nature Singapore Pte Ltd.
The registered company address is: 152 Beach Road, #21-01/04 Gateway East, Singapore 189721, Singapore

Preface

With the development of human society, the world is in urgent need of reducing the dependence on petroleum resources because its over-consumption will lead to a host of energy crisis issues and environmental problems, and utilization of renewable resources to replace fossil fuels has become extremely important. Over the past several decades, biomass as an abundant, renewable, less pollution and ecofriendly organic carbon resource to produce chemicals and biofuels has been proved a promising method and is currently receiving increasing attention of many chemists. The conversion of bio-derived platform chemicals from biomass is not only a hot topic, but also a common strategy to utilize biomass both in academia and industry.

The book covers advances on catalytic conversion of biomass and derivatives into useful chemicals and biofuels. It describes our recent researches relating to the hydrogenation of biomass and derivatives by diverse hydrogen sources such as water, isopropanol, gaseous hydrogen and $NaBH_4$ as well as their interesting mechanism aspects. A wide range of biomass and derivatives and some novel hydrogenation processes are involved in this book, and these works provide a wide ideas in novel hydrogenation of biomass and derivatives. Development strategies and challenges in the future research are also discussed.

This book will help readers to understand the hydrogenation of biomass and derivatives to useful chemicals and biofuels and also expand their knowledge of the utilization of renewable resources.

Shanghai, China Zhibao Huo

Contents

Part I Gaseous Hydrogen as Hydrogen Source

1 Selective Hydrogenation of Levulinate Esters to 1,4-Pentanediol Using a Ternary Skeletal CuAlZn Catalyst 3
1.1 Introduction 3
1.2 Experimental 5
1.2.1 Experimental Materials 5
1.2.2 Catalyst Preparation and Characterization 5
1.2.3 Typical Experiment and Product Analysis 5
1.3 Results and Discussion 6
1.4 Conclusions 14
References 15

Part II Isopropanol as Hydrogen Source

2 Catalytic Transfer Hydrogenation of Levulinate Ester into γ-Valerolactone Over Ternary $Cu/ZnO/Al_2O_3$ Catalyst 19
2.1 Introduction 19
2.2 Experimental 20
2.2.1 Experimental Materials 20
2.2.2 Catalyst Preparation 21
2.2.3 Catalyst Characterization 21
2.2.4 Typical Experiment and Product Analysis 22
2.3 Results and Discussion 22
2.3.1 Catalysts Characterization 22
2.3.2 Catalytic Activity of Various Catalysts for the EL Conversion 28
2.3.3 The Effect of the Various Parameters 32
2.3.4 Reusability of the CZA-3 Catalyst 34
2.3.5 Scopes of H-Donors and Raw Materials 36

2.3.6 The Proposed Mechanism of CTH Reaction of EL Over CZA-3 ... 37
2.4 Conclusions ... 38
References ... 39

3 Catalytic Transfer Hydrogenation of Ethyl Levulinate into γ-Valerolactone Over Air-Stable Skeletal Cobalt Catalyst ... 43
3.1 Introduction ... 43
3.2 Experimental ... 44
3.2.1 Materials ... 44
3.2.2 Pretreated Process for SCo ... 44
3.2.3 Catalyst Characterization ... 45
3.2.4 Experimental Procedures ... 46
3.3 Results and Discussion ... 47
3.3.1 Characterizations of Fresh-SCo, V-SCo, A-SCo and S-SCo ... 47
3.3.2 Catalysts Performance for the EL Conversion ... 50
3.3.3 The Effect of Various Parameters ... 53
3.4 The Availability of Air-Pretreated Process and the Stability of SCo Catalytic Activity ... 56
3.5 The Mechanism of EL CTH Process Over A-SCo ... 58
3.6 Conclusions ... 58
References ... 59

4 Boosting In-Situ Hydrodeoxygenation of Fatty Acids Over a Fine and Oxygen-Vacancy-Rich NiAl Catalyst ... 63
4.1 Introduction ... 63
4.2 Materials and Methods ... 65
4.2.1 Experimental Materials ... 65
4.2.2 Catalyst Preparation ... 65
4.2.3 Typical Experiment and Product Analysis ... 65
4.2.4 Density Functional Theory Calculations ... 66
4.3 Results and Discussion ... 67
4.3.1 Synthesis and Physicochemical Characterizations ... 67
4.4 Conclusions ... 80
References ... 81

Part III Water as Hydrogen Source

5 Chemoselective Synthesis of Propionic Acid from Biomass and Lactic Acid Over a Cobalt Catalyst in Aqueous Media ... 87
5.1 Introduction ... 87
5.2 Materials and Methods ... 89
5.2.1 Materials ... 89
5.2.2 Experimental Procedure ... 90
5.2.3 Products Analysis ... 90

5.3 Results and Discussion 91
5.3.1 Catalyst Screening 91
5.3.2 Effect of Reductant on the Yield of PA 91
5.3.3 Influence of Parameters on the Conversion of LA 95
5.3.4 Recyclability of Co Catalyst 98
5.3.5 Role of in Situ ZnO 99
5.3.6 Scope of Biomass Carbohydrates for PA Production 101
5.4 Conclusions 101
References 102

6 A Novel Approach for 1,2-Propylene Glycol Production from Biomass-Derived Lactic Acid 105
6.1 Introduction 105
6.2 Material and Methods 107
6.2.1 General Information 107
6.2.2 Product Analysis 107
6.2.3 Experimental Procedure for the Synthesis of 1,2-Propylene Glycol from Lactic Acid 108
6.3 Results and Discussion 108
6.3.1 Catalyst Screening 108
6.3.2 Optimization of Reductants 112
6.3.3 Effect of Various Parameters 113
6.3.4 Investigation of Gases Hydrogen 116
6.4 Mechanism 116
6.5 Conclusions 117
References 117

7 Catalytic Conversion of Ethyl Lactate to 1,2-Propanediol Over CuO 121
7.1 Introduction 121
7.2 Experimental 122
7.2.1 Experimental Materials 122
7.2.2 Experimental Procedure 123
7.2.3 Product Analysis 123
7.3 Results and Discussion 124
7.3.1 Catalyst Screening 124
7.3.2 The Optimization of the Various Parameters 125
7.3.3 The Formation of 1,2-PDO from Ethyl Lactate Using Gaseous Hydrogen 129
7.3.4 Mechanistic Studies 130
7.4 Conclusions 131
References 131

8 Highly Selective Hydrothermal Production of Cyclohexanol from Biomass-Derived Cyclohexanone Over Cu Powder 133
8.1 Introduction 133
8.2 Experimental 134
8.2.1 Experimental Materials 134
8.2.2 Experimental Procedure 135
8.2.3 Product Analysis 136
8.3 Results and Discussion 136
8.3.1 Catalyst Screening 136
8.3.2 Optimization of Active Metals 140
8.3.3 Effects of the Various Parameters 141
8.3.4 Cyclohexanone Hydrogenation by Added Hydrogen 144
8.3.5 Plausible Mechanism 144
8.4 Conclusions 145
References 145

9 Efficient Conversion of Dimethyl Phthalate to Phthalide Over CuO in Aqueous Media 147
9.1 Introduction 147
9.2 Experimental 148
9.2.1 Experimental Materials 148
9.2.2 Experimental Procedure 149
9.2.3 Product Analysis 150
9.3 Results and Discussion 150
9.3.1 Catalyst Screening 150
9.3.2 Effect of the Reductants 152
9.3.3 Effect of Other Parameters 155
9.3.4 Investigation of Gases Hydrogen 157
9.4 Mechanism Study 157
9.5 Conclusions 160
References 160

10 Heterogeneous Cu_2O-Mediated Ethylene Glycol Production from Dimethyl Oxalate 163
10.1 Introduction 163
10.2 Experimental 164
10.2.1 Materials and Methods 164
10.3 Results and Discussion 165
10.3.1 Metals or Metal Oxides Screening 166
10.3.2 Effect of Reductant 168
10.3.3 Effects of Water Filling, Reaction Temperature and Time 169
10.3.4 Investigation of Active Species 170
10.3.5 Recycle of Cu Formed in Situ 171
10.3.6 Mass Balance 172

10.4 Proposed Mechanism 172
10.5 Conclusions 173
References 173

11 Highly Efficient Conversion of Biomass-Derived Glycolide to Ethylene Glycol Over CuO in Water 177
11.1 Introduction 177
11.2 Experimental 178
11.2.1 General Information 178
11.2.2 Product Analysis 179
11.2.3 General Procedure for the Synthesis of Ethylene Glycol from Glycolide 179
11.2.4 General Procedure for the Synthesis of 1,2-Propanediol from DL-Lactide 180
11.3 Results and Discussion 180
11.4 Conclusions 186
References 186

Part IV Sodium Borohydride as Hydrogen Source

12 A Supported Ni Catalyst Produced from Ni–Al Hydrotalcite-Like Precursor for Reduction of Furfuryl Alcohol to Tetrahydrofurfuryl Alcohol by $NaBH_4$ in Water 191
12.1 Introduction 191
12.2 Results and Discussion 192
12.3 Conclusions 203
References 203

About the Author

Zhibao Huo is Distinguished Professor in College of Marine Ecology and Environment at Shanghai Ocean University, China. He earned his M.S. and Ph.D. from Tohoku University, Japan, and then went on to complete Postdoctoral Fellow. From 2011 to 2018, he went back to China and was promoted to work at Shanghai Jiao Tong University and also Endowed Professor sponsored by Special Appointment (Eastern Scholar) at Shanghai Institutions of Higher Learning. In 2018, he moved to Shanghai Ocean University as Professor. His research focuses on the development of novel catalysts for conversion of CO_2 and biomass into high-valued fuels and chemicals. He has authored more than 100 scientific publications, including peer-reviewed papers, patents and book chapters.

Part I
Gaseous Hydrogen as Hydrogen Source

Chapter 1
Selective Hydrogenation of Levulinate Esters to 1,4-Pentanediol Using a Ternary Skeletal CuAlZn Catalyst

1.1 Introduction

Urged by global energy crisis and environmental problems, numerous green technologies to utilize renewable resources have been developed for several decades. Biomass as an abundant, ecofriendly organic carbon resource is usually utilized to replace fossil fuels to produce chemicals and biofuels. Research on conversion of bio-derived platform chemicals from biomass is not only a hot topic, but also a common strategy to utilize biomass both in academia and industry (Gallezot, 2012, Alonso et al., 2010; Corma et al., 2007; Alonso et al., 2013; Behrens et al., 2012; Christian et al., 1947; Corbel-Demailly et al., 2013; Du et al., 2012; Fouilloux, 1983; Geilen et al., 2010, 2011; Gilkey & Xu, 2016; Hamminga et al., 2004; Herrmann & Emig, 1998; Horváth et al., 2008; Kang et al., 2015; Korstanje et al., 2015; Kuld et al., 2016; Li et al., 2016, 2014a, 2014b, 2015; Nakagawa et al., 2015; Phanopoulos et al., 2015; Pileidis & Titirici, 2016; Ren et al., 2016; Twigg & Spencer, 2001; vom Stein et al., 2014; Wang et al., 2014; Westerhaus et al., 2013; Xiao et al., 2015; Xu et al., 2014, 2016; Yang et al., 2013; Yoshino, 1989; Zhu et al., 2005). Because of the important commercial value of diols in biodegradable polyesters and organic synthesis industries, various bio-derived chemicals such as glycolide, oxalic acid, lactic acid, succinic acid, furfuryl alcohol, etc. were studied to produce diols (Kang et al., 2015; Korstanje et al., 2015; Li et al., 2016; Liu et al., 2015; Nakagawa et al., 2015; Xiao et al., 2015; Xu et al., 2014).

Levulinic acid (LA) with carbonyl and carboxyl groups in its carbon chain is one of the most important bio-derived platform molecules, which can be obtained from carbohydrate in aqueous solution (Pileidis & Titirici, 2016). Catalytic conversion of LA or levulinate esters (LEs) to 1,4-PDO is also viable in theory. Nowadays, due to the preferable intramolecular esterification of LA, most researches focus on hydrogenation of LA or LEs into γ-valerolactone (GVL), another versatile chemical (Alonso et al., 2013; Horváth et al., 2008). However, reports on the further hydrogenation of GVL into 1,4-PDO are few. It might be due to the high stable of the lactone

Z. Huo, *Diverse Hydrogen Sources for Biomass-derivatives Conversion*,
https://doi.org/10.1007/978-981-99-1673-3_1

ring structure in GVL to hinder the reaction, which makes the further hydrogenation process difficult and requires new catalytic system for the transformation.

In the early stage, copper-chromium oxide was investigated to catalyze LA and ethyl levulinate (EL) to produce 1,4-PDO (Christian et al., 1947). However, the system with harsh reaction condition (over 250 °C and at least 20 MPa hydrogen pressure), low yield and existing toxic metal Cr which raised the environmental risks. Recently, several researches focused on the designing homogeneous Ru complexes which showed excellent performances (Phanopoulos et al., 2015; vom Stein et al., 2014; Li et al., 2014b; Westerhaus et al., 2013; Geilen et al., 2010, 2011). Although the yield of 1,4-PDO was high under mild reaction conditions, expensive ligands and difficulty reusable characteristic for their catalysts cannot be avoided. Compared with homogeneous catalytic systems, heterogeneous catalysts, to some extent, can overcome above deficiencies. Hence, several bimetallic catalysts such as Ru-Re/C (Corbel-Demailly et al., 2013), Rh-Mo/SiO_2 (Li et al., 2014a), Ir-MoOx/SiO_2 (Wang et al., 2014) were developed to catalytic hydrogenation of LA or LEs. However, the low selectivity for production of 1,4-PDO (42.3–82%) and utilization of noble metals increased the cost. In the context, non-noble metal catalyst such as Cu was prepared to try to direct hydrogenate LA to 1,4-PDO (Xu et al., 2016). Unfortunately, the yield of 1,4-PDO was 22% even in a higher temperature (200 °C). Therefore, based on the previous research as we mentioned above, although some good results were reported, the highest yield of 1,4-PDO obtained from LA or LEs was 82% so far in heterogeneous catalytic system (Corbel-Demailly et al., 2013), conversion of LA or LEs to 1,4-PDO catalyzed by high efficient non-noble metal heterogeneous catalyst is still highly desirable and is a big challenge.

Skeletal Cu with high activity in the application of hydrogenation and dehydrogenation are wide used in chemical industry. Even more, the catalyst is low cost, simple preparation and good resistance to catalytic poisoning.

Because of its versatile properties, we here, prepared a ternary skeletal CuAlZn catalyst for hydrogenation of EL to 1,4-PDO. The 98% yield of 1,4-PDO was obtained (Fig. 1.1). To the best of our knowledge, it is the highest yield in heterogeneous catalysts so far.

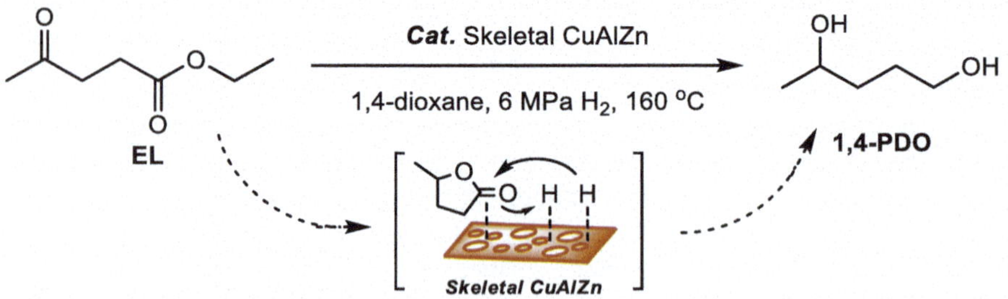

Fig. 1.1 CuAlZn-catalyzed hydrogenation of EL to 1,4-pentanediol. Reprinted from Royal Society of Chemistry (Ren et al., 2016)

1.2 Experimental

1.2.1 Experimental Materials

The materials in the study such as CuAlZn alloy, Cu, CuO, NaOH, ethyl levulinate and other organic solvents were all purchased from Sinopharm Chemical Reagent (China). 1,4-pentanediol was obtained from Sigma-Aldrich Co., Ltd. Pd/C (5 wt%) was purchased from Aladdin Chemical Reagent. CuAl alloy, Raney Ni, Raney Co, Raney Fe were purchased from Dalian Tongyong Chemical Co., Ltd.

1.2.2 Catalyst Preparation and Characterization

CuAlZn alloy powder (80–100 mesh) obtained by milling was treated with 20 wt% NaOH aqueous solution. Leaching process was conducted in an ice bath under the temperature range from 5 to 10 °C with vigorous stirring. After the desired leaching time, the powder was washed with deionized water until the pH was neutral and stored in ethanol.

XRD analysis was conducted in LabX XRD-6100 (Shimadzu) using a CuKα radiation source, the scanning speed was 6° min^{-1} and 2θ ranges from 10 to 80°. The surface area of catalyst was measured from N_2 adsorption–desorption isotherms by using automated gas sorption analyzer (Autosorb-iQ-TPX, Quantachrome, USA) with BET (Brunauer–Emmett–Teller) method. The element composition of alloy was analyzed by ICP-AES (inductively coupled plasma atomic emission spectroscopy, ICP-PS3500DD, HITACHI) after the alloy catalysts were digested with concentrated hydrochloric acid and nitric acid at first.

1.2.3 Typical Experiment and Product Analysis

All the experiments were conducted into a high pressure reactor with a 100 mL Teflon liner (PARR instrument company, USA). Desired amount of skeletal catalysts were taken out from ethanol and blow-dried with nitrogen gas. After feedstock, catalyst and solvent were loaded. The reactor was sealed and purged 5 times with hydrogen, and then desired hydrogen pressure was filled. After reaction, the reactor was cooled to room temperature. The liquid sample was collected, filtered with a 0.45 μm membrane and analyzed by GC-FID (gas chromatography with flame ionization detector, Agilent GC7890A), GC-MS (gas chromatography with mass spectrometer, Agilent GC7890A-MS5975C), which equipped with HP-Innowax column (30 m × 0.25 mm × 0.25 μm). The solid catalysts were recovered and washed with water and ethanol for several times, then kept into ethanol.

1.3 Results and Discussion

Commonly, commercial skeletal copper is prepared by CuAl alloy. Hence, the activities of CuAl alloy and its skeletal CuAl catalysts treated with alkali solution in different leaching time were first investigated (Table 1.1, entries 1–3). The leaching catalyst was named with the alloy and its leaching time. For example, the CuAlZn alloy which was treated with NaOH solution for 3.5 h could be named as CuAlZn-3.5 h. The GVL yields increased from 2.4 to 70.2% with the extending leaching time of CuAl alloy, indicating that longer leaching time presented higher activity for the hydrogenation of EL. But the selectivity of 1,4-PDO was still low. BET and ICP data showed that longer leaching time gave the larger specific surface area and higher content of Cu in alloy (Table 1.2, entries 1–3; Fig. 1.2a–c), which exposed more copper atoms onto catalyst surface from CuAl alloy phase and increased the active sites for the hydrogenation reaction. XRD patterns showed that $CuAl_2$ phase was gradually weakened and the broad Cu phase peaks appeared during the leaching process of CuAl alloy (Fig. 1.3). And the poor results were also given by CuZn alloy and its leaching catalyst (Table 1.1, entries 4–5). The low activity of CuZn-3.5 h catalyst may attribute to the hard leaching process of CuZn alloy under present condition, which could not lead the alloy to form the skeletal structure. It is proved by the results that no obviously differences were observed from XRD patterns, and the ratio of Cu/Zn and BET data on the CuZn alloy before and after treatment with NaOH solution (Figs. 1.2d, e, 1.3; Table 1.2, entries 4–5). BET surface area of CuZn-3.5 h was only 6.8 m^2/g (Table 1.2, entry 5; Fig. 1.2e), which was much smaller than skeletal CuAl and CuAlZn (Table 1.2, entries 2–3, 7–10; Fig. 1.2g–j).

Table 1.1 Hydrogenation of EL with various catalysts[a]

Entry	Catalyst	Conv. (%)	Yield (%)	
			GVL	1,4-PDO
1	CuAl-alloy	2.5	2.4	0
2	CuAl-0.5 h	47.7	38.6	9
3	CuAl-3.5 h	71.2	70.2	< 1
4	CuZn-alloy	3.6	3.5	0
5	CuZn-3.5 h	4.5	4.2	0
6	CuAlZn-alloy	3	3	0
7	CuAlZn-1.5 h	100	61.7	38.2
8	CuAlZn-2.5 h	100	55.6	44.3
9	CuAlZn-3.5 h	100	17.4	82.6
10	CuAlZn-4 h	100	18.2	81.7
11[b]	CuAlZn-3.5 h	100	17.0	82.9

[a] Reaction condition: 1 mmol EL, 50 mg catalysts, 10 mL dioxane, 160 °C, 4 MPa H_2, 10 h; [b] The catalyst was kept in ethanol for 1 week before the activity test. Reprinted from Royal Society of Chemistry (Ren et al., 2016)

Table 1.2 Characteristic of skeletal copper-based catalysts

Entry	Catalyst	BET surface area ($m^2\ g^{-1}$)	Elementary composition Cu/Al/Zn (wt%)
1	CuAl-alloy	6.4	46.5/53.5/0
2	CuAl-0.5 h	28.1	77.5/22.5/0
3	CuAl-3.5 h	37.5	91.0/8.9/0
4	CuZn-alloy	2.1	53.0/0/47.0
5	CuZn-3.5 h	6.8	53.9/0/46.2
6	CuAlZn-alloy	2.4	52.3/46.0/1.6
7	CuAlZn-1.5 h	36.5	77.4/20.5/2.1
8	CuAlZn-2.5 h	40.5	83.8/13.9/2.2
9	CuAlZn-3.5 h	48.1	89.0/8.6/2.3
10	CuAlZn-4 h	55.8	91.0/7.0/2.0

Reprinted from Royal Society of Chemistry (Ren et al., 2016)

Indeed, Zn is a vital active promoter for Cu-based catalyst to hydrogenate lactone ring structure into diols (Hamminga et al., 2004; Herrmann & Emig, 1998), a ternary CuAlZn alloy containing only 1.6 wt% Zn was used for preparing skeletal catalysts (Table 1.2, entry 6).

TEM images show that CuAlZn alloy treated by alkali possess clear nanoporous structure, which increasing the surface area of catalyst (Fig. 1.4a, b). Although the alloy of CuAlZn almost has no effect for EL conversion (Table 1.1, entry 6), the leaching one showed the great performance for both the EL conversion and diol production (Table 1.1, entry 7), which far surpassed the skeletal CuAl, though both of them contained similar BET surface areas and the Cu content (Table 1.1, 2–1, 2, 3, 7). The products were confirmed by GC-MS and GC-FID (Fig. 1.5, 5). The same as the skeletal CuAl, leaching time is a remarkable effect on the catalytic activity of skeletal CuAlZn (Table 1.1, entries 6–10). With the leaching process prolonging, the content of Cu increased and Al reduced, but the ratio of Zn was stable in the catalyst (Table 1.3, entries 6–10). The leaching process monitored by XRD showed that CuAl2 phases disappeared and Cu phases arose, but no Zn or its alloy phases were observed, maybe due to its low amount (Fig. 1.6). The highest yield of 1,4-PDO is up to 82.6% when alkali treatment time for the catalyst is 3.5 h (Table 1.1, entry 9). It indeed seems that small amount of additive Zn enhances the selectivity of 1,4-PDO as we have previously assumed. The mechanism of the lactone ring-opening is still obscure, it may attribute to the higher oxophilicity of Zn in comparison to Cu, which gives rise to the species with oxygen atoms to bind readily with Zn and leads to the formation of an oxidation state of Zn (Behrens et al., 2012; Kuld et al., 2016), and then the active Zn-bonding intermediate could be ring opened to form diol. Longer treatment time with alkaline solution did not make the catalyst CuAlZn-4 h to exhibit higher activity (Table 1.1, entry 10), even the catalyst possessed larger BET surface area, higher Cu content and low amount of Al than CuAlZn-3.5 h (Table 1.3, entry 10;

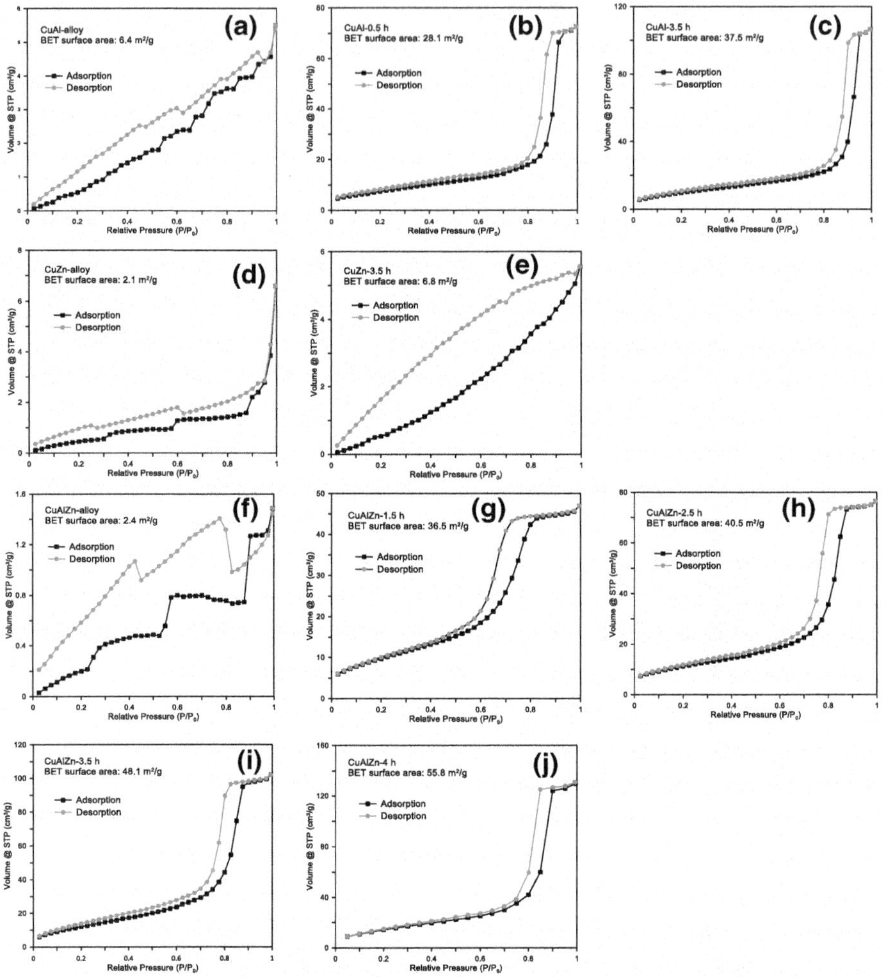

Fig. 1.2 **a** Nitrogen isotherms of CuAl alloy; **b** Nitrogen isotherms of CuAl-0.5 h; **c** Nitrogen isotherms of CuAl-3.5 h; **d** Nitrogen isotherms of CuZn alloy; **e** Nitrogen isotherms of CuZn-3.5 h; **f** Nitrogen isotherms of CuAlZn alloy; **g** Nitrogen isotherms of CuAlZn-1.5 h; **h** Nitrogen isotherms of CuAlZn-2.5 h; **i** Nitrogen isotherms of CuAlZn-3.5 h; **j** Nitrogen isotherms of CuAlZn-4 h. Reprinted from Royal Society of Chemistry (Ren et al., 2016)

Fig. 1.2j). In addition, keeping a certain amount of Al in the skeletal catalyst could enhance the activity (Gilkey & Xu, 2016; Yang et al., 2013), because the residual Al could increase the interstitial lattice defects in the skeletal particles (Fouilloux, 1983; Yoshino, 1989). The catalyst lifetime was also tested. Kept in ethanol for one week before use, the skeletal CuAlZn-3.5 h gave a 82.9% yield of 1,4-PDO without the loss of activity (Table 1.1, entry 11). Hot filtration and leaching tests were carried out and no metal ions (Cu, Al, Zn) were detected.

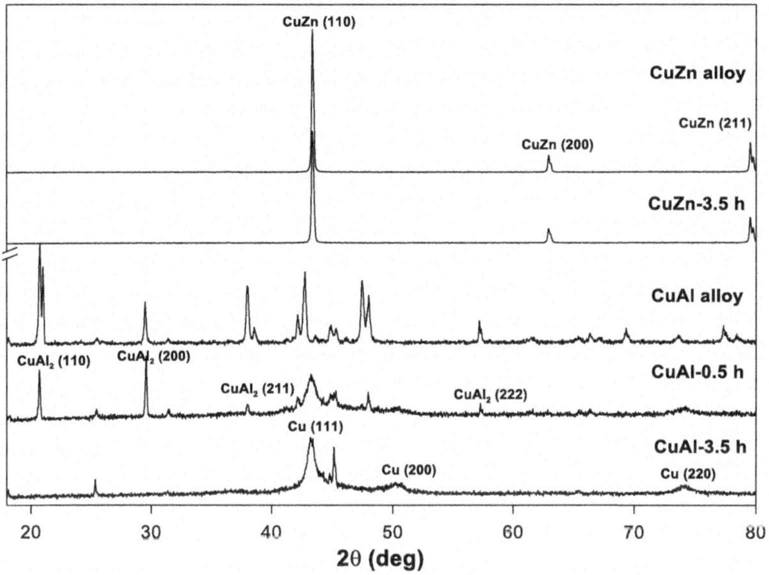

Fig. 1.3 XRD patterns for leaching process of CuAl and CuZn alloys. Reprinted from Royal Society of Chemistry (Ren et al., 2016)

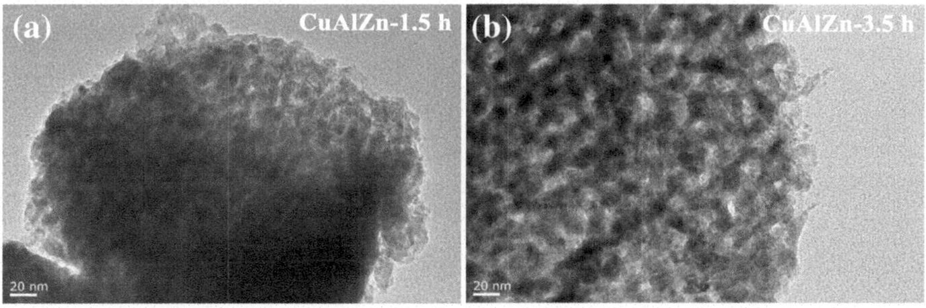

Fig. 1.4 **a** TEM image of CuAlZn-1.5 h; **b** TEM image of CuAlZn-3.5 h. Reprinted from Royal Society of Chemistry (Ren et al., 2016)

In addition, other Cu species such as Cu and CuO were also tested. As is expected, no 1,4-PDO were observed (Table 1.4, entries 1–2). Since the skeletal metal catalysts have good catalytic ability for hydrogenation of EL, skeletal Ni, Co, Fe were also evaluated. However, skeletal Ni and Co given the high yield of GVL, the selectivities of 1,4-PDO were still very low (Table 1.4, entries 3–5). Noble-metal-based catalysts such as Pd/C did not display 1,4-PDO selectivity (Table 1.4, entry 6).

Considering the greatly influence of solvents for catalytic reaction, some of commonly used solvents were tested as shown in Table 1.3. Polar 1,4-dioxane was first investigated and gave a good result, the 1,4-PDO yield was 82.6% (entry 1).

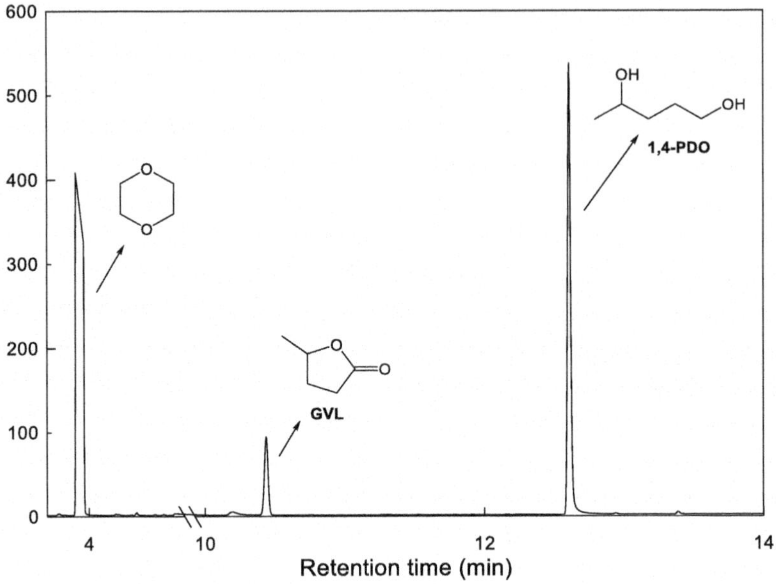

Fig. 1.5 GC-FID for hydrogenation of EL (1 mmol EL, 50 mg CuAlZn-3.5 h, 10 mL dioxane, 160 °C, 4 MPa H_2, 10 h). Reprinted from Royal Society of Chemistry (Ren et al., 2016)

Table 1.3 Various solvents for hydrogenation of EL

Entry	Solvent	Conv. (%)	Yield (%)	
			GVL	1,4-PDO
1	1,4-dioxane	100	17.4	82.6
2	Water	78.8	73.8	0
3	EtOH	100	69.8	30.1
4	i-PrOH	88.7	80.1	6.0
5	THF	100	33.3	66.6
6	DMF	90.5	86.4	1.3
7	DMSO	5	0	0

Reaction condition: 1 mmol EL, 50 mg CuAlZn-3.5 h, 10 mL solvent, 160 °C, 4 MPa H_2, 10 h (EtOH: ethanol, i-PrOH: propanol, THF: tetrahydrofuran, DMF: N,N-Dimethylfor-mamide, DMSO: dimethyl sulfoxide). Reprinted from Royal Society of Chemistry (Ren et al., 2016)

Hence, several polar solvents were also screened. It was disappointed that the selectivities of 1,4-PDO were low, although water, EtOH, i-PrOH and THF afforded the good conversion of EL (entries 2–5), even they were reported as suitable solvents for diols production (Corbel-Demailly et al., 2013; Du et al., 2012; Li et al., 2014a, 2014b; Phanopoulos et al., 2015; Wang et al., 2014; Xu et al., 2016). DMF gave the 86.4% yield of GVL, but 1.3% yield of 1,4-PDO (entry 6). And DMSO totally

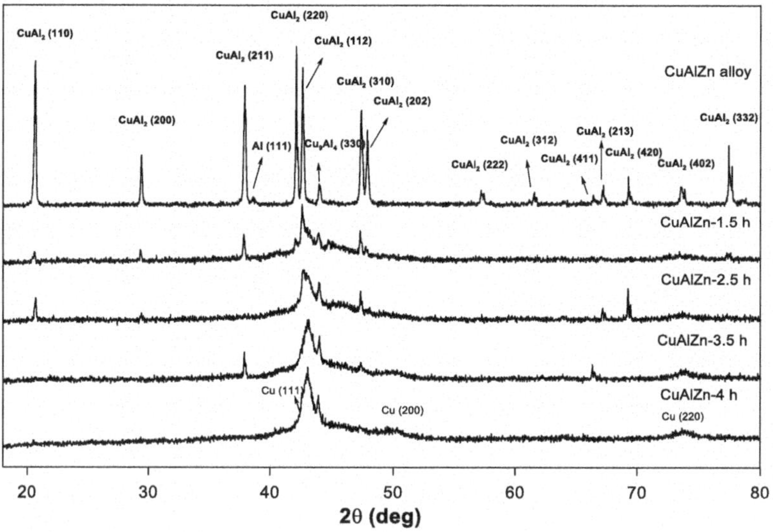

Fig. 1.6 XRD patterns for leaching process of CuAlZn alloy. Reprinted from Royal Society of Chemistry (Ren et al., 2016)

Table 1.4 Hydrogenation of EL into 1,4-PDO with various catalysts

Entry	Catalyst	Conv. (%)	Yield (%)	
			GVL	1,4-PDO
1	Cu	17.9	17.9	0
2	CuO	26.6	25.5	0
3	Raney Ni	92.0	84.2	3.8
4	Raney Co	100	97.2	2.4
5	Raney Fe	21.4	21.3	0
6	Pd/C	16.5	16.5	0

Reaction condition: 1 mmol EL, 50 mg catalysts, 10 mL dioxane, 160 °C, 4 MPa H_2, 10 h. Reprinted from Royal Society of Chemistry (Ren et al., 2016)

inhibited the hydrogenation of EL, no desired products were observed (entry 7). This might be the S species in DMSO poisoned the active sites on the catalyst surface (Twigg & Spencer, 2001). Thus, 1,4-dioxane was defined as suitable solvent for next investigation.

Catalyst loadings were investigated for the conversion of EL, and the results were summarized in Fig. 1.7a. The skeletal CuAlZn exhibited great performance, even when only 10 mg catalyst was used. The conversion rate of EL was 100% and the yields of GVL and 1,4-PDO were 72.6 and 27.3%, respectively. Improving catalyst loading obviously increased the 1,4-PDO selectivity. The highest 1,4-PDO yield was

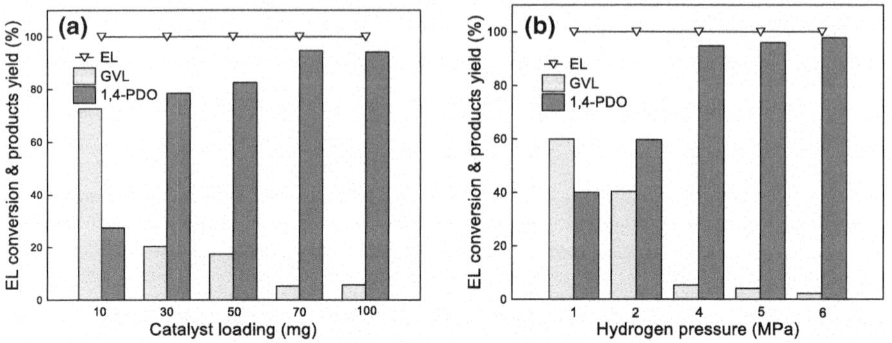

Fig. 1.7 **a** The effect of skeletal CuAlZn-3.5 h catalyst loading (1 mmol EL, 10 mL 1,4-dioxane, 160 °C, 4 MPa H_2, 10 h); **b** The effect of hydrogen pressure (1 mmol EL, 70 mg CuAlZn-3.5 h, 10 mL 1,4-dioxane, 160 °C, 10 h). Reprinted from Royal Society of Chemistry (Ren et al., 2016)

94.7% when 70 mg catalyst was loaded. Over the amount of 70 mg, the yield cannot be up to a higher level.

H_2 pressure experiments were conducted to investigate the hydrogenation of EL over skeletal CuAlZn-3.5 h catalyst in 1,4-dioxane (Fig. 1.7b). It can be seen that higher pressure is preferable for GVL opening its lactone ring structure to produce 1,4-PDO. 1 MPa H_2 gave the only 40% yield of 1,4-PDO and 59.9% GVL was remained. When H_2 pressure exceeded 4 MPa, the yield of 1,4-PDO was more than 90%. The maximum yield was up to 97.7% when 6 MPa H_2 was afforded.

Temperature also shows the greatly influence on the selectivity of 1,4-PDO production (see Table 1.5). Although all of EL was converted at the investigated temperature, hydrogenation of GVL to 1,4-PDO was more difficult. At low temperature such as 120 °C, only 24.5% yield of 1,4-PDO was obtained after 10 h (entry 1). Therefore, high reaction temperature is necessary for the hydrogenation of GVL to 1,4-PDO. The best reaction temperature was 160 °C to achieve the 97.7% yield of 1,4-PDO (entry 2). Interestingly, over 160 °C, no more GVL was converted, even 1,4-PDO yield declined slightly (entries 3–4), which might be attributed to the acceleration of its reverse reaction under the high temperature. Previous research also reported the dehydrogenation of 1,4-butanediol to γ-butyrolactone over a CuAlZn catalyst under H_2 atmosphere (Zhu et al., 2005). To confirm the hypothesis, an evidential experiment was conducted using 1,4-PDO as feedstock. 1 mmol 1,4-PDO was hydrogenated over 70 mg CuAlZn-3.5 h with 6 MPa H_2 under 200 °C for 5 h, the result showed that 10.9% 1,4-PDO was converted to obtain 7.3% yield of GVL.

The time course of the conversion of EL catalyzed by skeletal CuAlZn-3.5 h under 160 °C with 6 MPa H_2 pressure was investigated. As shown in Fig. 1.8, the CuAlZn-3.5 h displayed excellent performance for EL conversion, the EL was almost exhausted within 30 min, and 55.6% yield of GVL and 44% yield of 1,4-PDO were obtained, respectively. In comparison to the skeletal CuAl catalyst, which only 71.2% of EL was converted after 10 h reaction (Table 1.1, entry 3), the additive small amount of Zn (around 2 wt%) into the CuAl alloy could effectively enhance the conversion

Table 1.5 The effect of reaction temperature for hydrogenation of EL

Entry	Temp (°C)	Conv. (%)	Yield (%)	
			GVL	1,4-PDO
1	120	100	75.4	24.5
2	160	100	2.2	97.7
3	180	100	4.5	95.5
4	200	100	8.2	91.5

Reaction condition: 1 mmol EL, 70 mg CuAlZn-3.5 h, 10 mL 1,4-dioxane, 6 MPa H_2, 10 h. Reprinted from Royal Society of Chemistry (Ren et al., 2016)

rate of EL hydrogenation. It seems that the GVL synthesis from EL was readily. But the ring opening of GVL to 1,4-PDO was a little bit difficult, 3 h was good time for 1,4-PDO yield of up to 96%. After 3 h, only slightly increase was observed. The highest yield of 1,4-PDO was 98% when the time was prolonged to 6 h.

According to our findings and previous researches (Hamminga et al., 2004; Herrmann & Emig, 1998; Xiao et al., 2015; Xu et al., 2014, 2016), a possible pathway of conversion of EL into 1,4-PDO over skeletal CuAlZn catalyst was proposed in Fig. 1.9. No intermediates were detected during the hydrogenation reaction. The highly active skeletal CuAlZn extremely accelerate the EL conversion, no EL could be detected even if the reaction was carried out for only 30 min. γ-hydroxypentanoic

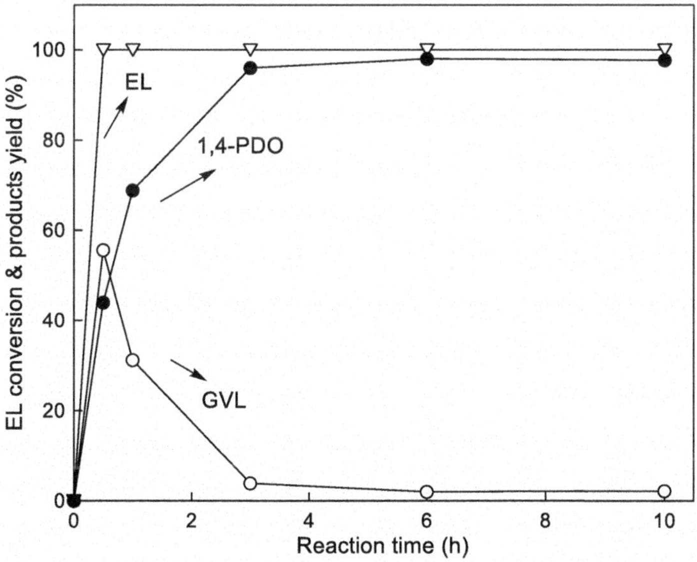

Fig. 1.8 The time course of conversion of EL (1 mmol EL, 70 mg CuAlZn-3.5 h, 10 mL 1,4-dioxane, 160 °C, 6 MPa H_2). Reprinted from Royal Society of Chemistry (Ren et al., 2016)

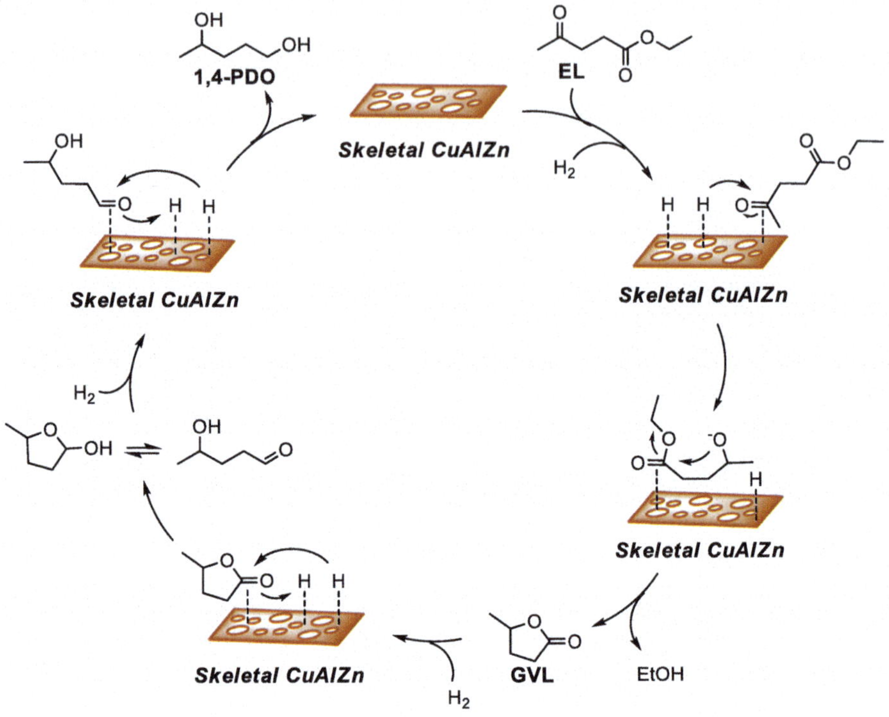

Fig. 1.9 The proposed pathways for the hydrogenation of EL to 1,4-PDO catalyzed by skeletal CuAlZn. Reprinted from Royal Society of Chemistry (Ren et al., 2016)

ester, a unstable compound, is a possible hydrogenation intermediate for GVL formation (Xu et al., 2016). After that, the lactone ring adsorbs on the catalyst, with the promotion of Zn on alloy surface, the C=O is hydrogenated to C–OH. Then the formed 2-hydroxy-5-methyltetrahydrofuran is ring opened to form 4-hydroxypentanal. The final product 1,4-PDO was obtained by the hydrogenation of 4-hydroxypentanal along with the regeneration of skeletal CuAlZn.

1.4 Conclusions

In summary, we developed a highly efficiency non-noble skeletal CuAlZn catalysts for conversion of EL to 1,4-PDO. 98% yield of 1,4-PDO was obtained within 6 h at 6 MPa H_2 and 160 °C. It is the highest yield according to previous research in heterogeneous catalytic system till now. Compared with three kinds of Cu-based alloy for prepared skeletal catalysts, the imperative role of Zn for 1,4-PDO production not only greatly enhance the conversion of EL to GVL, but also accelerate the lactone ring-opening process to produce diol. Several parameters such as catalyst loading, solvents, H_2 pressure and reaction time were also the important factors for the reaction proceeding. The present study provides a mild, low cost, high efficient approach for 1,4-PDO synthesis from bio-based levulinic esters.

References

Alonso, D. M., Bond, J. Q., & Dumesic, J. A. (2010). Catalytic conversion of biomass to biofuels. *Green Chemistry, 12*, 1493–1514.

Alonso, D. M., Wettstein, S. G., & Dumesic, J. A. (2013). Gamma-valerolactone, a sustainable platform molecule derived from lignocellulosic biomass. *Green Chemistry, 15*, 584–596.

Behrens, M., Studt, F., Kasatkin, I., Kuhl, S., Havecker, M., Abild-Pedersen, F., Zander, S., Girgsdies, F., Kurr, P., Kniep, B. L., Tovar, M., Fischer, R. W., Norskov, J. K., & Schlogl, R. (2012). The active site of methanol synthesis over $Cu/ZnO/Al_2O_3$ industrial catalysts. *Science, 336*, 893–897.

Christian, R. V., Brown, H. D., & Hixon, R. M. (1947). Derivatives of γ-Valerolactone, 1,4-Pentanediol and 1,4-Di-(β-cyanoethoxy)-pentane1. *Journal of the American Chemical Society, 69*, 1961–1963.

Corbel-Demailly, L., Ly, B. K., Minh, D. P., Tapin, B., Especel, C., Epron, F., Cabiac, A., Guillon, E., Besson, M., & Pinel, C. (2013). Heterogeneous catalytic hydrogenation of biobased levulinic and succinic acids in aqueous solutions. *ChemSusChem, 6*, 2388–2395.

Corma, A., Iborra, S., & Velty, A. (2007). Chemical routes for the transformation of biomass into chemicals. *Chemical Reviews, 107*, 2411–2502.

Du, X.-L., Bi, Q.-Y., Liu, Y.-M., Cao, Y., He, H.-Y., & Fan, K.-N. (2012). Tunable copper-catalyzed chemoselective hydrogenolysis of biomass-derived γ-valerolactone into 1,4-pentanediol or 2-methyltetrahydrofuran. *Green Chemistry, 14*, 935–940.

Fouilloux, P. (1983). The nature of raney nickel, its adsorbed hydrogen and its catalytic activity for hydrogenation reactions (review). *Applied Catalysis, 8*, 1–42.

Gallezot, P. (2012). Conversion of biomass to selected chemical products. *Chemical Society Reviews, 41*, 1538–1558.

Geilen, F. M. A., Engendahl, B., Harwardt, A., Marquardt, W., Klankermayer, J., & Leitner, W. (2010). Selective and flexible transformation of biomass-derived platform chemicals by a multifunctional catalytic system. *Angew, 122*, 5642–5646.

Geilen, F. M., Engendahl, B., Holscher, M., Klankermayer, J., & Leitner, W. (2011). Selective homogeneous hydrogenation of biogenic carboxylic acids with $[Ru(TriPhos)H]^+$: A mechanistic study. *Journal of the American Chemical Society, 133*, 14349–14358.

Gilkey, M. J., & Xu, B. (2016). Heterogeneous catalytic transfer hydrogenation as an effective pathway in biomass upgrading. *ACS Catalysis, 6*, 1420–1436.

Hamminga, G. M., Mul, G., & Moulijn, J. A. (2004). Real-time in situ ATR-FTIR analysis of the liquid phase hydrogenation of -butyrolactone over Cu-ZnO catalysts: A mechanistic study by varying lactone ring size. *Chemical Engineering Science, 59*, 5479–5485.

Herrmann, U., & Emig, G. (1998). Kinetics and mechanism in the liquid-phase hydrogenation of maleic anhydride and intermediates. *Chemical Engineering and Technology, 21*, 285–295.

Horváth, I. T., Mehdi, H., Fábos, V., Boda, L., & Mika, L. T. (2008). γ-Valerolactone—A sustainable liquid for energy and carbon-based chemicals. *Green Chemistry, 10*, 238–242.

Kang, K. H., Hong, U. G., Bang, Y., Choi, J. H., Kim, J. K., Lee, J. K., Han, S. J., & Song, I. K. (2015). Hydrogenation of succinic acid to 1,4-butanediol over Re–Ru bimetallic catalysts supported on mesoporous carbon. *Applied Catalysis a: General, 490*, 153–162.

Korstanje, T. J., Van Der Vlugt, J. I., Elsevier, C. J., & De Bruin, B. (2015). Hydrogenation of carboxylic acids with a homogeneous cobalt catalyst. *Science, 350*, 298–302.

Kuld, S., Thorhauge, M., Falsig, H., Elkjaer, C. F., Helveg, S., Chorkendorff, I., & Sehested, J. (2016). Quantifying the promotion of Cu catalysts by ZnO for methanol synthesis. *Science, 352*, 969–974.

Li, M., Li, G., Li, N., Wang, A., Dong, W., Wang, X., & Cong, Y. (2014a). Aqueous phase hydrogenation of levulinic acid to 1,4-pentanediol. *Chemical Communications, 50*, 1414–1416.

Li, W., Xie, J.-H., Yuan, M.-L., & Zhou, Q.-L. (2014b). Ruthenium complexes of tetradentate bipyridine ligands: Highly efficient catalysts for the hydrogenation of carboxylic esters and lactones. *Green Chemistry, 16*, 4081–4086.

Li, L., Ren, D., Fu, J., Liu, Y., Jin, F., & Huo, Z. (2016). Heterogeneous Cu_2O-mediated ethylene glycol production from dimethyl oxalate. *Journal of Energy Chemistry, 25*, 507–511.

Liu, X., Wang, X., Xu, G., Liu, Q., Mu, X., & Liu, H. (2015). Tuning the catalytic selectivity in biomass-derived succinic acid hydrogenation on FeOx-modified Pd catalysts. *Journal of Materials Chemistry A, 3*, 23560–23569.

Nakagawa, Y., Tamura, M., & Tomishige, K. (2015). Catalytic Conversions of Furfural to Pentanediols. *Catalysis Surveys from Asia, 19*, 249–256.

Phanopoulos, A., White, A. J. P., Long, N. J., & Miller, P. W. (2015). Catalytic transformation of Levulinic acid to 2-methyltetrahydrofuran using ruthenium–N-triphos complexes. *ACS Catalysis, 5*, 2500–2512.

Pileidis, F. D., & Titirici, M. M. (2016). Levulinic acid biorefineries: New challenges for efficient utilization of biomass. *Chemsuschem, 9*, 562–582.

Ren, D., Wan, X., Jin, F., Song, Z., Liu, Y., & Huo, Z. (2016). Selective hydrogenation of levulinate esters to 1,4-pentanediol using a ternary skeletal CuAlZn catalyst. *Green Chemistry, 18*, 5999–6003.

Twigg, M. V., & Spencer, M. S. (2001). Deactivation of supported copper metal catalysts for hydrogenation reactions. *Applied Catalysis a: General, 212*, 161–174.

Vom Stein, T., Meuresch, M., Limper, D., Schmitz, M., Holscher, M., Coetzee, J., Cole-Hamilton, D. J., Klankermayer, J., & Leitner, W. (2014). Highly versatile catalytic hydrogenation of carboxylic and carbonic acid derivatives using a Ru-triphos complex: Molecular control over selectivity and substrate scope. *Journal of the American Chemical Society, 136*, 13217–13225.

Wang, Z., Li, G., Liu, X., Huang, Y., Wang, A., Chu, W., Wang, X., & Li, N. (2014). Aqueous phase hydrogenation of acetic acid to ethanol over Ir-MoOx/SiO_2 catalyst. *Catalysis Communications, 43*, 38–41.

Westerhaus, F. A., Wendt, B., Dumrath, A., Wienhofer, G., Junge, K., & Beller, M. (2013). Ruthenium catalysts for hydrogenation of aromatic and aliphatic esters: Make use of bidentate carbene ligands. *Chemsuschem, 6*, 1001–1005.

Xiao, J., Huo, Z., Ren, D., Zhang, S., Luo, J., Yao, G., & Jin, F. (2015). A novel approach for 1,2-propylene glycol production from biomass-derived lactic acid. *Process Biochemistry, 50*, 793–798.

Xu, L., Huo, Z., Fu, J., & Jin, F. (2014). Highly efficient conversion of biomass-derived glycolide to ethylene glycol over CuO in water. *Chemical Communications, 50*, 6009–6012.

Xu, Q., Li, X., Pan, T., Yu, C., Deng, J., Guo, Q., & Fu, Y. (2016). Supported copper catalysts for highly efficient hydrogenation of biomass-derived levulinic acid and γ-valerolactone. *Green Chemistry, 18*, 1287–1294.

Yang, Z., Huang, Y. B., Guo, Q. X., & Fu, Y. (2013). RANEY® Ni catalyzed transfer hydrogenation of levulinate esters to gamma-valerolactone at room temperature. *Chemical Communications, 49*, 5328–5330.

Yoshino, T. (1989). Surface characteristics and activities of plate-type Raney nickel catalyst. *Journal of Catalysis, 118*, 436–442.

Zhu, Y.-L., Yang, J., Dong, G.-Q., Zheng, H.-Y., Zhang, H.-H., Xiang, H.-W., & Li, Y.-W. (2005). An environmentally benign route to γ-butyrolactone through the coupling of hydrogenation and dehydrogenation. *Applied Catalysis, B: Environmental, 57*, 183–190.

Part II
Isopropanol as Hydrogen Source

Chapter 2
Catalytic Transfer Hydrogenation of Levulinate Ester into γ-Valerolactone Over Ternary $Cu/ZnO/Al_2O_3$ Catalyst

2.1 Introduction

Nowadays, the depletion of fossil resources and increasing CO_2 emission problems pushes the scholars to focus on developing renewable biomass resources to reduce the dependence on fossil resources (De et al., 2015; Hu et al., 2017; Tang et al., 2017). Among bio-based chemicals, Levulinic acid (LA) and its esters, are considered as one of the most important lignocellulosic derived materials and a kind of versatile biomass platform molecules (Khan et al., 2018). γ-valerolactone (GVL), which can be obtained by the catalytic hydrogenation of LA, has been identified as one of the most alluring platform molecules which is a sustainable commodity chemical and has tremendous potential as a biofuel due to its combustion properties. Therefore, generating GVL from biomass has drawn an increasing attention.

Recently, catalytic transfer hydrogenation (CTH) process has been regarded as an attractive approach for energy-saving and cost-effective method in comparison to the former hydrogenation that needs high-pressure and flammable H_2 using precious metal catalysts (Wang et al., 2016; Zhang et al., 2017). Among several hydrogen donors used in the CTH process, 2-butanol (2-BuOH) (Luo et al., 2014) and i-propanol (*i*-PrOH) (Al-Shaal et al., 2016; Kuwahara et al., 2014b; Li et al., 2017; Yang et al., 2016) that belong to secondary alcohols are widely used because they are corrosion resistance, non-toxic and no greenhouse gas (CO_2) release compared to formic acid (Yuan et al., 2013) and methanol (Li et al., 2015; Tang et al., 2015).

Dumesic et al. firstly reported a viable and green process called CTH for quantitative yields of GVL using secondary alcohols as H-donor over heterogeneous catalysts ZrO_2 in 150 °C over 16 h in 2011 (Chia & Dumesic, 2011). After that, more and more Zr-based heterogeneous catalysts had been reported (Kuwahara et al., 2014a, 2017a, 2017b; Song et al., 2015; Tang et al., 2014). Although many noble and non-noble catalysts were studied to facilitate this hydrogenation process under H_2, there were very short of researches on catalyst use in the CTH process, only Ru (Al-Shaal et al., 2016; Kuwahara et al., 2014b), Pd (Amarasekara & Hasan, 2015), Ni (Hengne et al., 2016; Zhen et al., 2013) and Cu (Yang et al., 2016) were investigated. Nevertheless,

Z. Huo, *Diverse Hydrogen Sources for Biomass-derivatives Conversion*,
https://doi.org/10.1007/978-981-99-1673-3_2

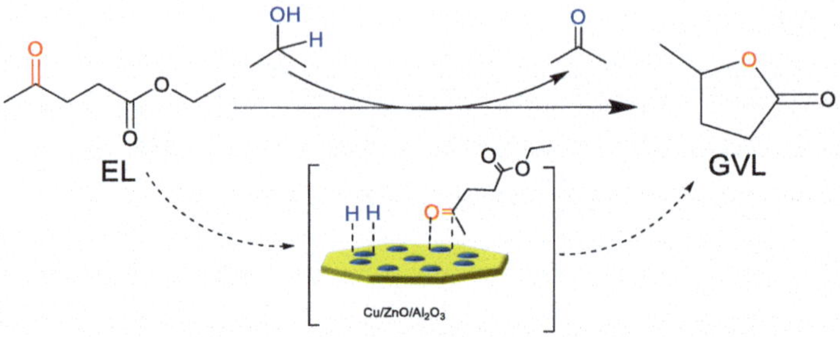

Fig. 2.1 CTH of EL into GVL over $Cu/ZnO/Al_2O_3$ catalyst. Reprinted from Elsevier (Zhang et al., 2019)

the use of costly Ru and Pd made against to the aim of producing low price liquid transportation fuels, and cheap Ni is a heavy metal with high toxicity that limits their large-scale application to a certain degree. Thus, to develop a low-cost and high-activity catalyst that can be industrial application is significant. Recently, nano copper had attracted research attention because of its excellent performance in the hydrogenation and dehydrogenation process. However, supported Cu catalysts were used rarely in this CTH process to produce GVL up to now. In 2016, Ge Tian et al. performed a CTH process of ML over $Cu\text{-}ZrO_2$ catalyst in which 75% yields of GVL was obtained, but the performance of its cyclic stability is bad because of inevitable leaching of copper particles (Yang et al., 2016). Hence, moderate GVL yield and selectivity and the difficulty of reusing impeded the catalyst application. To this end, it is urgent to explore a nano Cu-based catalyst with high activity and stability to solve problems mentioned above.

Herein, an effective ternary $Cu/ZnO/Al_2O_3$ catalyst which was prepared by two-step co-precipitation method and can be sustainable use was explored for CTH process of ethyl levulinate (EL) to produce GVL under mild condition with *i*-PrOH as hydrogen donor (Fig. 2.1). Comparing with the binary Cu/ZnO and Cu/Al_2O_3 catalysts, the influence of different support oxides was explored both in the catalytic performance and the structure of catalysts.

2.2 Experimental

2.2.1 Experimental Materials

EL (ethyl levulinate, 99%) as the initial reactant was purchased from Beijing inoke Technology Co. Ltd. AR grade of $Cu(NO_3)_2$, $Zn(NO_3)_2$, $Al(NO_3)_3$ and Na_2CO_3 were purchased from Sinopharm Chemical Reagent (China).

2.2.2 *Catalyst Preparation*

Ternary Cu/ZnO/Al_2O_3 (CZA) catalysts with Cu/Zn/Al molar ratio of 6:3:1 were prepared by the coprecipitation method on the basis of procedure reported by Baltes C (Baltes et al., 2008). In a general procedure, metal nitrates [($Cu(NO_3)_2$ (0.6 mol/L), $Zn(NO_3)_2$ (0.3 mol/L), $Al(NO_3)_3$ (0.1 mol/L))] were dissolved together as the corresponding solutions, and a mixed solution Na_2CO_3 (1 mol/L) was prepared as the precipitating agent. Typically, two solutions were dumped together into 200 mL of deionized water under continuously stirring and maintained at 70 °C with a constant pH (± 0.1 unit) = 7, controlled by Mettler-Toledo FiveEasy Plus™ (FE20). After the coprecipitation stopped and then aging for 1 h. After that, the obtained slurry was filtered and washed with deionized water, then dried overnight at 80 °C and finally calcined at 300 °C for 3 h in air after grounding to obtain CZA precursor. Ultimately, CZA catalyst was obtained when the precursors were reduced with a 5% H_2/Ar mixture at 300 °C for 3 h in a tubular furnace.

For comparison, binary Cu/ZnO (CZ) catalyst with Cu/Zn molar ratio of 6:4 and binary Cu/Al_2O_3 (CA) catalyst with Cu/Al molar ratio of 6:4 were prepared by the coprecipitation method mentioned above, respectively.

Considering the effect of calcination temperature on the catalyst activity, a series of CZA catalysts were prepared under different calcination temperature, such as CZA-3, CZA-4, CZA-5, CZA-6, CZA-7 were calcined at 300, 400, 500, 600, 700 °C, respectively.

2.2.3 *Catalyst Characterization*

Powder X-ray diffraction (XRD) patterns of the samples were obtained by LabX XRD-6100 X-ray Diffractometer with Cu Kα radiation ($\lambda = 1.54056$ nm) and the 2θ angle in the range of 10° to 80°at a speed of 6° min^{-1}. Brunauer–Emmett–Teller (BET) surface area and Barrett–Joyner–Halenda (BJH) mesopore and Horvath–Kawazoe (HK) microspore size distribution were determined by static nitrogen adsorption–desorption technique on a Autosorb-iQ-TPX, Quantachrome, USA after the catalysts were degassed at 120 °C under vacuum for at least 12 h. Scanning electron microscopy (SEM) characterizations combined with energy-dispersive X-ray spectroscopy (EDX) were taken on a */JEOL JSM-7800F Prime Super resolution field emission scanning electron microscope (ER-SEM) to elucidate the morphology and microstructure of the catalysts. Transmission electron microscopy (TEM) was done with */JEOL JEM-2100 transmission electron microscopy to observe the microscopic particle shape and size of catalysts. Hydrogen–temperature-programmed reduction (H_2-TPR) experiments were conducted on a Autosorb-iQ-TPX, Quantachrome, USA to determine the reducibility of different copper oxides of the catalysts. Liquid samples were collected and filtered with a 0.22 μm membrane, the following was to analyze by GC-FID (Gas chromatography with flame ionization detector, Agilent GC7890A) and GC-MS (Gas chromatography with mass spectrometer, Agilent GC7890A-MS5975C) equipped with HP-Innowax column (30 m $\times$ 0.25 mm $\times$ 0.25 μm).

2.2.4 *Typical Experiment and Product Analysis*

All the experiments were performed in a 5 mL glass vessel covered with a Teflon top that could bear high pressure. Firstly, feedstock, catalyst and solvent were loaded into the reactor, then a certain N_2 was filled in and the vessel was sealed. After that, it was put onto magnetic stirring apparatus at the desired reaction temperature for a certain time. Finally, after reaction, the reactor was cooled to room temperature, liquid samples were collected and filtered, and the following was to analyze by GC-FID and GC-MS. As for the solid samples, they were collected and washed with deionized water for several times for XRD analysis and then kept in ethanol for subsequent analysis.

2.3 Results and Discussion

2.3.1 *Catalysts Characterization*

XRD patterns of Cu particle with different support oxides were shown in Fig. 2.2a. The peaks of Cu were observed in the 2θ ranges of 42.0°–42.5°, 48.5°–52.0° and 73.5°–75.0° in all samples. The strong diffraction peaks of Cu particles were detected in CA catalysts, but no strong diffraction peaks of Al_2O_3 were detected which might be due to its high dispersion or amorphous phase (Igor et al., 2007). Both the diffraction peaks of Cu and ZnO particles were observed in CZ catalysts, whereas only the diffraction peaks of Cu was detectable when the Al_2O_3 was put together with Cu and ZnO as a composition of the CZA-3 catalysts. Comparison of particle size of Cu in CA, CZ and CZA-3 calculated by XRD diffraction patterns using Scherrer Equation, which were 18.5, 15.4 and 12.7 nm, respectively (Table 2.1). The changes in the crystallite sizes suggested a gradual decrease in particle sizes of Cu crystallite, in other words, the dispersion of metallic Cu was remarkably enhanced with ZnO and Al_2O_3 as binary support. And that was agreement with the conclusions in the literatures (Figueiredo et al., 1998; Shishido et al., 2006).

The structural properties of the different catalysts were shown in Table 2.1. The BET surface area of CZA-3, CZ and CA were 56.4, 40.6 and 114.9 m^2/g, respectively. The order of BET surface area was found to be CA > CZA-3 > CZ, the order of crystallite size of Cu was found to be CA > CZ > CZA-3. Based on the data of BET and crystallite size of Cu, it illustrated that the addition of Al_2O_3 might effectively improve the overall surface area of the catalyst, and the presence of ZnO might be

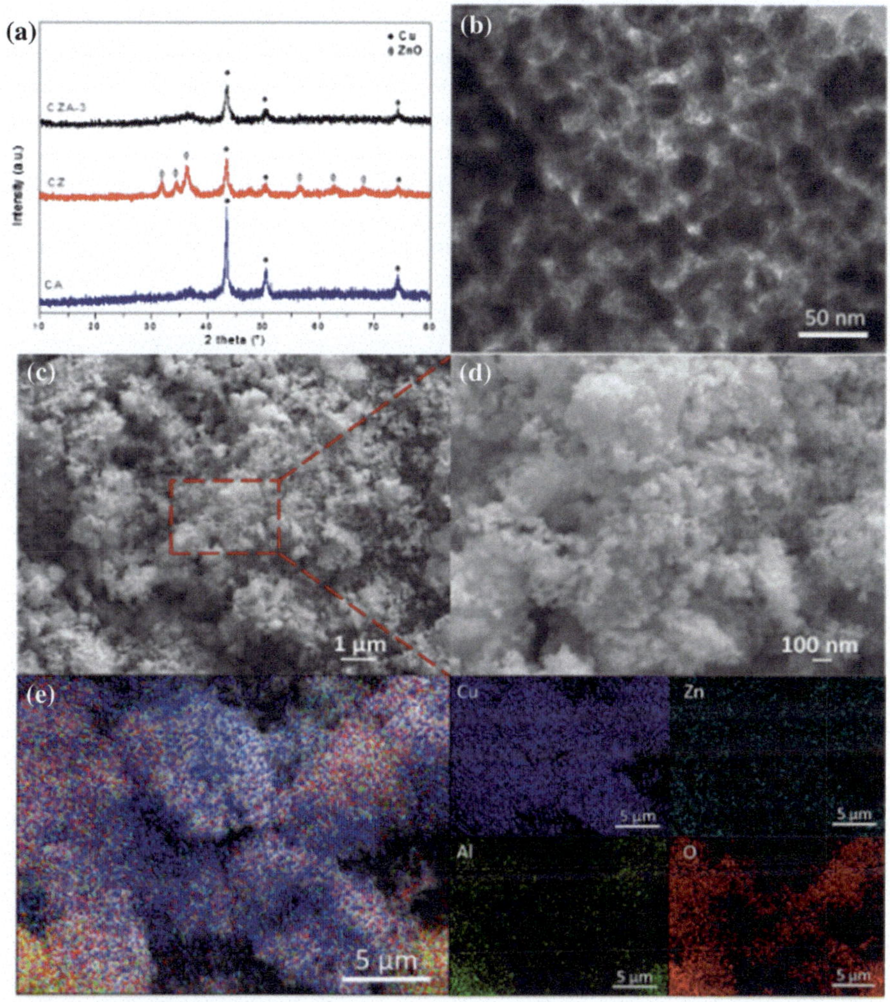

Fig. 2.2 **a** XRD patterns of nanocatalysts: CZA-3, CZ, CA; **b** TEM image of CZA-3; **c**, **d** SEM images; and **e** EDX mapping of CZA-3. Reprinted from Elsevier (Zhang et al., 2019)

Table 2.1 Structural properties of the different catalysts

Cat	Cu/Zn/Al (mol%)	N_2 adsorption/desorption			X-ray diffraction
		S_{BET}[a] (m^2/g)	V_p[b] (cc/g)	d_p[c] (nm)	d_{Cu}[d] (nm)
CZA-3	60/30/10	56.4	0.454	3.408	12.7
CZ	60/40/0	40.6	0.418	3.058	15.4
CA	60/0/40	114.9	0.582	3.053	18.5

Reprinted from Elsevier (Zhang et al., 2019)

[a] BET surface area. [b] Total pore volume determined as the N_2 adsorption. [c] Average pore diameter.
[d] Cu mean crystallite size determined by X-ray diffraction

helpful to get Cu particles with smaller crystallite size. Therefore, it seemed that Al_2O_3 and ZnO effectively regulated the specific surface area and the crystallite size of Cu of the catalysts.

The SEM images of CZA-3 catalysts were shown in Fig. 3.2c, d, EDX and dot-mapping micrographs of synthesized nanocatalysts were displayed in Fig. 2.2e. Generally, small particles were observed in Fig. 2.2d. In addition, nothing particularly different between CZ sample (see Fig. 3.3a, b), CA sample (see Fig. 3.3c, d) and CZA-3 could be found from the SEM images. The EDX mapping micrograph of CZA-3 in Fig. 2.2e exhibited uniform dispersion of Cu, Zn, Al and O elements. The EDX mapping micrographs of CZ (see Fig. 2.3g) and CA (see Fig. 2.3h) exhibited uniform dispersion of Cu, Zn, O and Cu, Al, O elements, respectively.

From TEM micrographs, it was obvious that the morphology and particle size of CZA-3 (see Fig. 2.2b), CZ (see Fig. 2.3e) and CA (see Fig. 2.3f) were different. CZA-3 had the uniform particle size mostly about 20 nm. Figure 2.3e showed the microstructure of CZ which had the size from 10 to 30 nm. As shown in Fig. 2.3f, the particles of CA were not well shaped and several particles also stuck together. Thus, CZA-3 had the more uniform and homogeneous particles than CZ and CA (Igor et al., 2007).

The HRTEM images of CZA-3 was presented in Fig. 2.4. The interplanar distances were measured and annotated in the corresponding position. The Cu particles were in contact with several ZnO particles. It seemed that ZnO particles serve as spacers between Cu particles which probably prevented Cu particles from sintering (Igor et al., 2007). Because the Al_2O_3 was amorphous, there was no clear lattice stripe displayed, it might be in the place where the stripes were unclear (Fichtl et al., 2015; Guo et al., 2009; Igor et al., 2007).

The results of H_2-TPR measurements were shown in Fig. 2.5. According to the literature, the reduction of CZAO-3 could proceed in three steps (Bahmani et al., 2016; Ren et al., 2015). The curves of CZA-3 and CZ were deconvoluted into 3 reduction peaks by the Guassian-type function. The three different intensity bands are at 170–200 °C, 200–220 °C and 220–250 °C, respectively. That could correspond to the reduction of three kinds of Cu^{2+} into Cu, such as highly dispersed CuO particles (peaks 1 and 4), bulk CuO with ZnO contact (peaks 2 and 4) and bulk CuO without ZnO contact (peaks 3 and 6). Comparing with the high-temperature peak represented bulk CuO without ZnO contact, peak 3 of CZAO-3 was much lower than peak 6 of CZO. It could be speculated that the addition of Al_2O_3 increases the interaction between CuO and ZnO (Atake et al., 2007). On the other hand, the curves of CA were more complicated and they were deconvoluted into five reduction peaks, while the peaks 7 and peak 8 correspond to highly dispersed CuO particles and bulk CuO with Al_2O_3 contact, respectively. And the peaks of temperature > 250 °C (peaks 9, 10 and 11) could be large CuO particles (López-Suárez et al., 2008). It was obvious that CAO was much more difficult to be reduced without the addition of ZnO. Therefore, ZnO/Al_2O_3 as a binary support might intense the metal-support interaction of CZA-3 and that might have an influence on the enhanced reducibility of copper (Bahmani et al., 2016; Zhang et al., 2014).

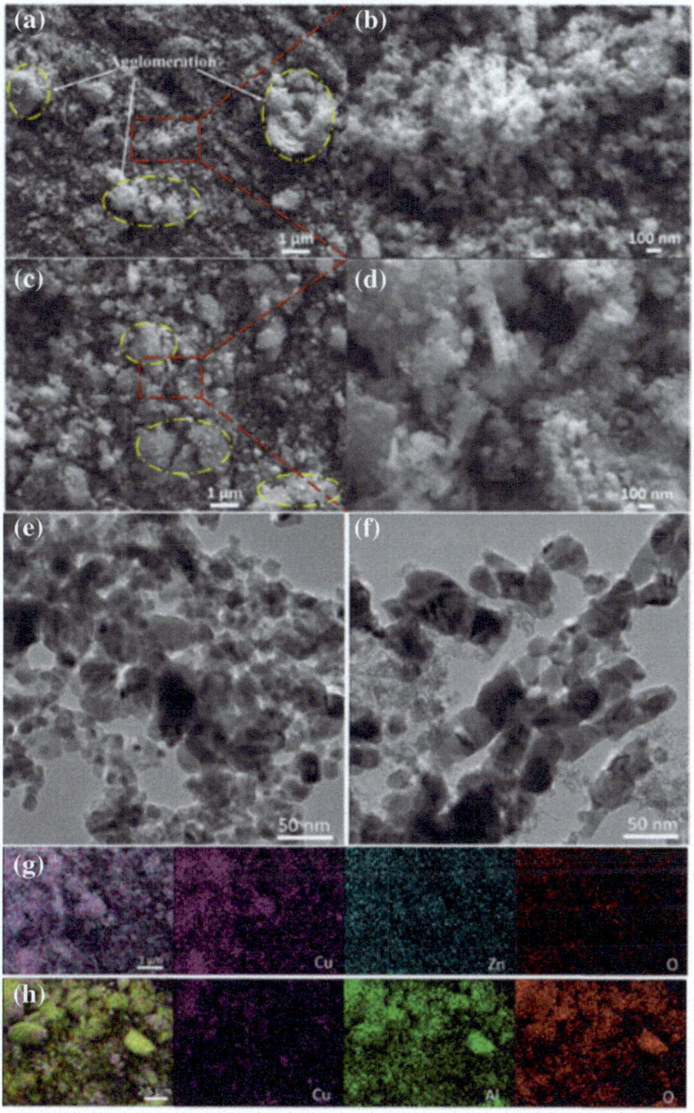

Fig. 2.3 **a**, **b** SEM images of CZ; **c**, **d** SEM images of CA; **e** TEM images of CZ; **f** TEM images of CA; **g** EDX mapping of CZ; **h** EDX mapping of CA. Reprinted from Elsevier (Zhang et al., 2019)

Calcination temperature was essentially important for the formation of nondecomposed Cu/Zn hydroxy carbonate residues in the calcined precursors (Baltes et al., 2008), which had strong influence on the Cu NPs and the metal-support interaction after H_2 reduction (Zhang et al., 2014). Therefore, a series of CZA catalysts under different calcination temperature were prepared to investigate the effect on the morphologies and the catalytic activities of the finally resulted samples.

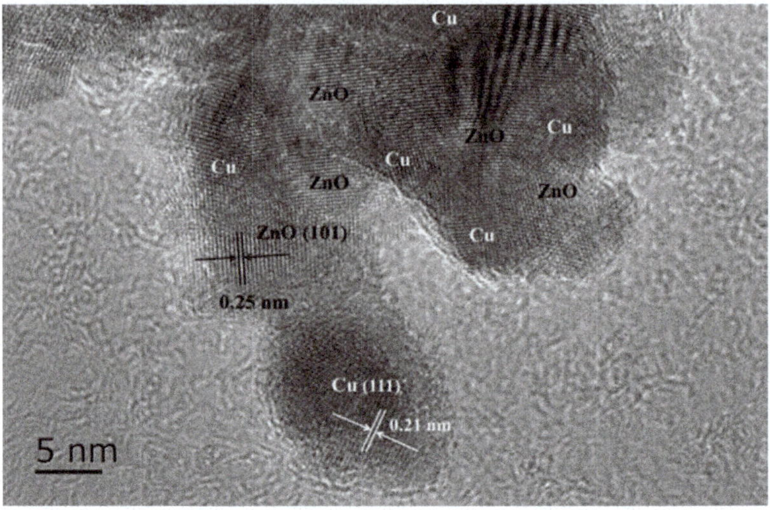

Fig. 2.4 HRTEM image of CZA-3. Reprinted from Elsevier (Zhang et al., 2019)

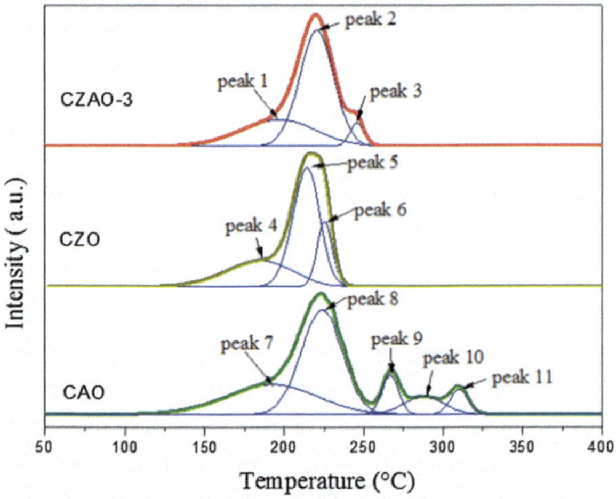

Fig. 2.5 TPR profiles for catalysts divided by means of deconvolution method. Operating conditions: 40 ml/min (STP), H_2 (5 vol.%)/Ar, Sample weight, 50 mg. Thick curved showed experimental plot of TPR profiles. Solid thin lines were deconvoluted curves. Reprinted from Elsevier (Zhang et al., 2019)

XRD patterns of a series of CZA catalysts calcined at different temperature were shown in Fig. 2.6. The peaks of Cu were observed in all samples, and the intensity of peaks increased with the increase of calcination temperature, especially for the CZA-7 sample, which meant the higher crystallinity and the sintering of Cu particles. Besides, the peaks of ZnO catalysts were not obvious in CZA-3, CZA-4 and CZA-5, which illustrated the low crystallinity of ZnO. However, the peaks of ZnO could be observed prominently in CZA-6 and CZA-7, which indicated that high temperature could promote crystallization of ZnO particles. As for Al_2O_3, no diffraction peaks were exhibited in all samples, indicating its amorphous nature and/or high dispersion

(Igor et al., 2007). The Cu crystallite size of CZA-3 and CZA-7 calculated by XRD diffraction patterns using Scherrer Equation were 12.7 and 23.5 nm, respectively. It was consistent with the crystallite size and BET surface area results of corresponding CZAO precursors (see Table 2.2), the CuO crystallite size of CZAO-3 and CZAO-7 were 5.8 and 20.4 nm, respectively. On the other hand, with the increase of calcination temperature, the BET surface area of catalyst precursors had been decreasing from 125.5 m^2/g (CZAO-3) to 26.7 m^2/g (CZAO-7). Apparently, CZAO-3 had the smallest CuO crystallite size and the largest BET surface area. Besides, since the surface area largely decreased to about 26 m^2/g in CZAO-7, it seemed that high calcination temperature caused the collapse of porous structure of catalyst.

From the SEM images of CZA catalysts in Fig. 2.7, it seemed that bigger cluster could be seen in CZA-7 catalyst. The EDX mapping micrograph of CZA catalysts in

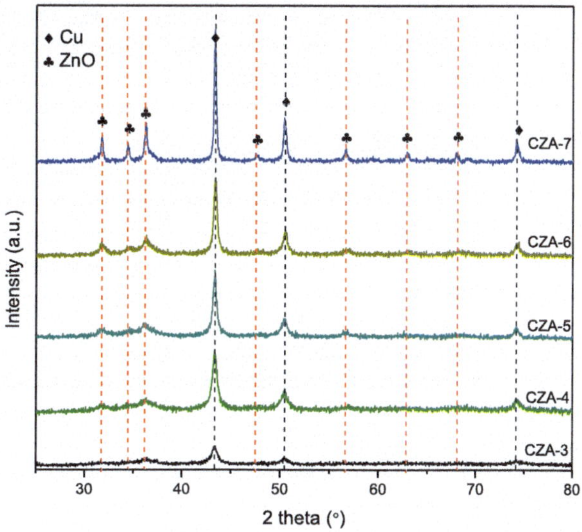

Fig. 2.6 XRD patterns of synthesized nanocatalysts calcined at different temperature (300–700 °C). Reprinted from Elsevier (Zhang et al., 2019)

Table 2.2 Structural properties of a series of CZAO precursors

Catalyst	N_2 adsorption/desorption			X-ray diffraction
	S_{BET}[a] (m^2/g)	V_p[b] (cc/g)	d_p[c] (nm)	d_{CuO}[d] (nm)
CZAO-3	125.5	0.657	7.775	5.8
CZAO-4	60.5	0.525	5.584	7.5
CZAO-5	57.3	0.511	3.055	9.1
CZAO-6	45.8	0.412	3.828	13.3
CZAO-7	26.7	0.237	4.892	20.4

Reprinted from Elsevier (Zhang et al., 2019)
[a] BET surface area. [b] Total pore volume determined as the N_2 adsorption. [c] Average pore diameter. [d] CuO mean crystallite size as determined by X-ray diffraction

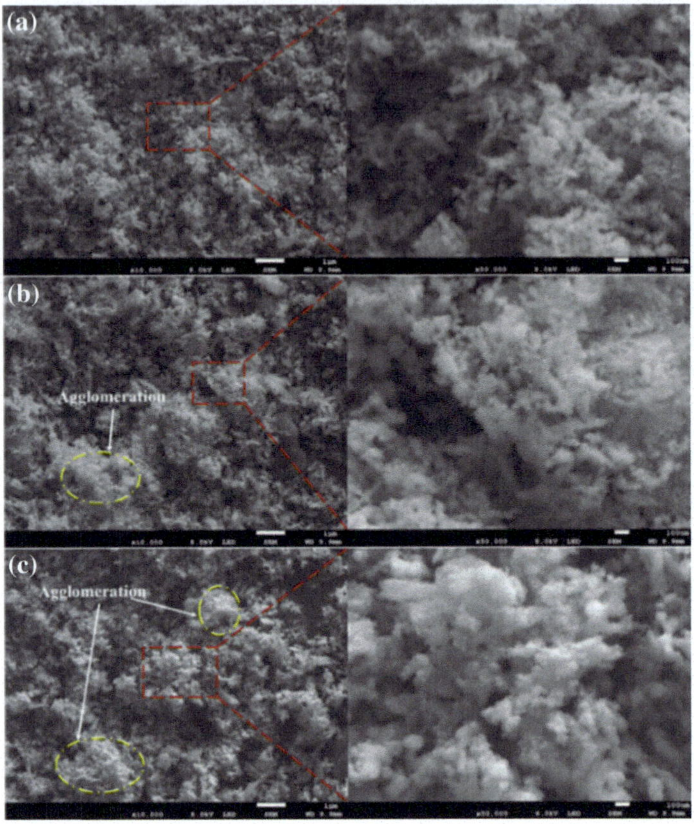

Fig. 2.7 SEM images of **a** CZA-5, **b** CZA-6 and **c** CZA-7. Reprinted from Elsevier (Zhang et al., 2019)

Fig. 2.8 exhibited uniform dispersion of Cu, Zn, Al and O elements and there were no particular difference could be found. The photographs of TEM indicated that CZA-6 and CZA-7 had larger particle size (see Fig. 2.9a, b), and had much different morphology compared to CZA-3. Regarding to above descriptions, it can be assumed that high temperature calcination results in sintering and recrystallization of copper and the collapse of porous structure.

2.3.2 Catalytic Activity of Various Catalysts for the EL Conversion

The catalytic activity of CZA-3, CZ, CA and other catalysts were characterized for CTH process as shown in Table 2.3. In the initial tests, the blank test showed that it was not possible to covert EL to GVL using *i*-PrOH as an H-donor without any

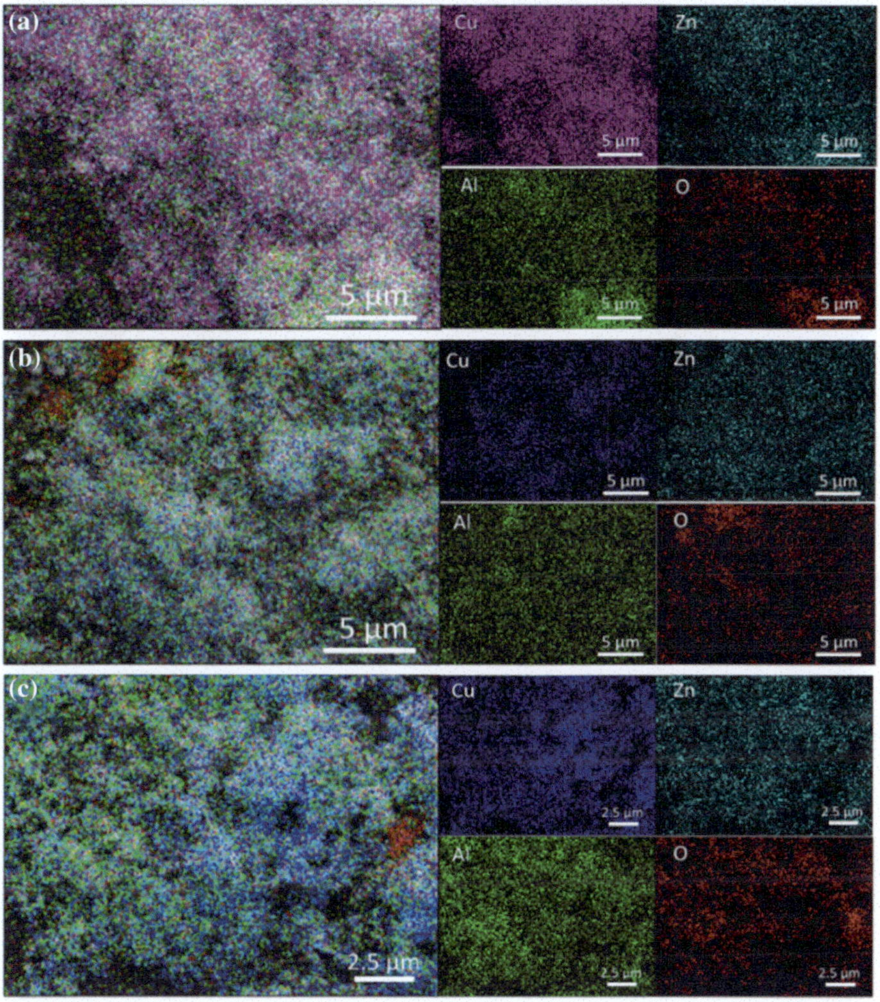

Fig. 2.8 EDX mappings of **a** CZA-5, **b** CZA-6 and **c** CZA-7. Reprinted from Elsevier (Zhang et al., 2019)

catalysts (entry 1). Notably, 90.7% conversion of EL and 88.6% yield of GVL were observed when CZA-3 as a catalyst was used (entry 2).

The yields of GVL were 82.4 and 58.9% over CZ and CA, respectively. Compared the results of CZ with CA, the EL conversion of 90.5% over CZ was much higher than 65.7% over CA (entries 3 and 4). Since the crystallite sizes of Cu in CA, CZ were 18.5 and 15.4 nm, respectively, it seemed that the smaller crystallite size of Cu was a reason for the higher activity of CZ to facilitate the conversion of EL. On the other hand, compared the results of CZA-3 with CZ (entries 2 and 3), the EL conversion was almost the same, but the GVL yield of 88.6% over CZA-3 was higher than

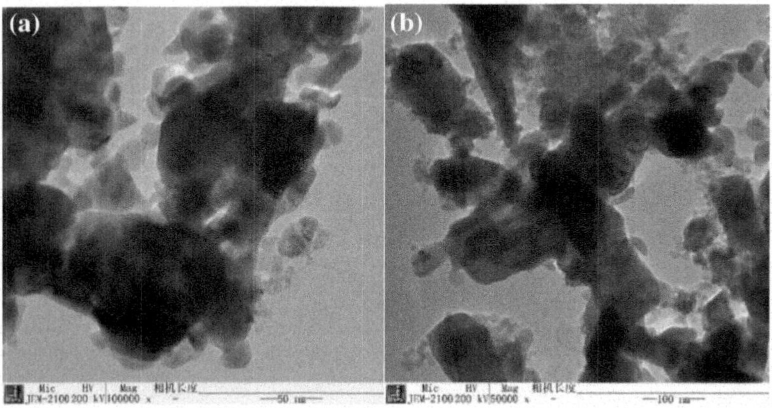

Fig. 2.9 TEM images of **a** CZA-6, **b** CZA-7. Reprinted from Elsevier (Zhang et al., 2019)

Table 2.3 Catalyst screening for GVL production from EL. Reprinted from Elsevier (Zhang et al., 2019)

Entry	Cat	Conv. (%)	Sel. (%)	GVL yield (%)
1	Blank	0	0	0
2	CZA-3	90.7	97.7	88.6
3	CZ	90.5	91.0	82.4
4	CA	65.7	89.6	58.9
5	ZnO	1.2	0	0
6	Al_2O_3	0.8	0	0
7	$Cu/ZnO/ZrO_2$	49.5	69.1	34.2
8	Skeletal Cu	90.6	72.7	65.9
9	CZA-4	90.6	95.4	86.4
10	CZA-5	89.3	96.0	85.8
11	CZA-6	84.3	92.9	78.3
12	CZA-7	2.5	56.0	1.4

Conditions 0.2 mmol EL, 20 mg catalysts, 2 mL *i*-PrOH, 120 °C, 3 h

82.4% over CZ. Besides, it could be seen that CZA-3 showed the superior catalytic activity and GVL selectivity of 97.7% was obtained compared to the catalysts of CZ (91%) and CA (89.6%). In addition, GC-FID photographs of solutions catalyzed by CZA-3 and CZ were shown in Fig. 2.10, the peaks of EL of two catalysts were almost the same, but the peak of GVL of CZA-3 was a little higher, and another peak at retention time ranging from 10.0 to 10.1 min which might be an intermediate was much smaller than that of CZ. It seemed that the addition of Al_2O_3 made CZA-3 own the smaller crystallite size of 12.7 nm and the highest dispersion of metallic Cu

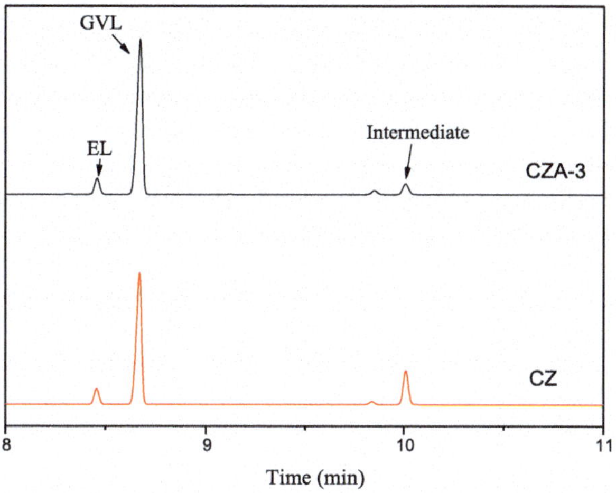

Fig. 2.10 GC-FID photographs of solutions catalyzed by CZA-3 and CZ, reaction conditions: 0.2 mmol EL, 20 mg catalyst, 2 mL *i*-PrOH, 120 °C, 3 h. Reprinted from Elsevier (Zhang et al., 2019)

mentioned before, which probably increased the activity of CZA-3 by accelerating the transformation of intermediate to GVL.

To highlight the best performance of ternary catalyst of CZA-3, other catalysts such as ZnO, Al_2O_3, $Cu/ZnO/ZrO_2$ and skeletal Cu were also examined (entries 5–8). The results illustrated that single ZnO or Al_2O_3 did not exhibit catalytic activity for the transformation. However, all of CZA-3, CZ and CA with ZnO and/or Al_2O_3 as support had catalytic activity, indicating that the active species in the catalysts was metallic Cu. The use of $Cu/ZnO/ZrO_2$ gave 34.2% yield (entry 7) and skeletal Cu afforded 65.9% yield (entry 8). Several other Cu-based catalysts such as nanoporous copper catalyst CuNPore, Cu powder, CuO, Cu/ZrO_2, Cu/SiO_2, CuAlZn-alloy and Cu/MgO were also tested, however, no GVL was detected. In addition, nanoporous nickel catalyst NiNPore and Pd/C which had been widely used in the hydrogenation field were attempted, however it seemed that they could not trigger this reaction. Therefore, CZA-3 was chosen to follow the optimization to obtain higher yield of GVL.

To highlight the best performance of ternary catalysts CZA-3, other catalysts such as ZnO, Al_2O_3, $Cu/ZnO/ZrO_2$ and skeletal Cu were also examined (entries 5–8). The results illustrated that single ZnO or Al_2O_3 did not exhibit catalytic activity for the transformation. However, all of CZA-3, CZ and CA with ZnO and/or Al_2O_3 as support had catalytic activity, indicating that the active species in the catalysts was metallic Cu. The use of $Cu/ZnO/ZrO_2$ gave 34.2% yield (entry 7) and Skeletal Cu afforded 65.9% yield (entry 8). Several other Cu-based catalysts such as nanoporous copper catalyst CuNPore, Cu power, CuO, Cu/ZrO_2, Cu/SiO_2, CuAlZn-alloy and Cu/MgO were also tested, however, no GVL was detected. In addition, nanoporous nickel catalyst NiNPore and Pd/C which had been widely used in the hydrogenation field were attempted, however it seemed that they could not trigger this reaction. Therefore, CZA-3 was chosen to follow the optimization to obtain higher yield of GVL.

It was reported that synthesized CZA catalysts from different calcination temperature having different morphological structure, the surface properties and the active phase dispersion directly affected catalytic activity (Kowalik & Próchniak, 2010). Thus, a series of CZA catalysts including CZA-4, CZA-5, CZA-6 and CZA-7 prepared in different calcination temperature were tested for the CTH reaction (entries 9–12). As a result, the yield of GVL was significantly affected by CZA catalysts in different calcination temperature. Obviously, CZA-3 had the best performance compared to others. The catalytic activity of CZA-4, CZA-5 and CZA-6 decreased a little than CZA-3, but CZA-7 as a catalyst only gave 1.4% yield of GVL (entry 12). Based on the BET results and crystallite size in Table 2.2, it assumed that the larger size of Cu crystallites and the collapse of porous structure might be two main reasons for CZA-7 deactivation at the high calcination temperature.

2.3.3 *The Effect of the Various Parameters*

To clarify the effect of CZA-3 loading for CTH reaction, the experiments were carried out as shown in Fig. 2.11. It revealed that the conversion of EL increased from 70.0 to 97.9% when CZA-3 loading increased from 10 to 60 mg, while the yield of GVL up to 92.65% was obtained with 30 mg. However, only a small extent of increase of 1.65% when CZA-3 loading was 60 mg. Considering the economic problem, 30 mg of CZA-3 was used in further study.

Next, the influences of reaction time and temperature were investigated. The reaction was conducted with CZA-3 at 120 and 140 °C for different reaction time of

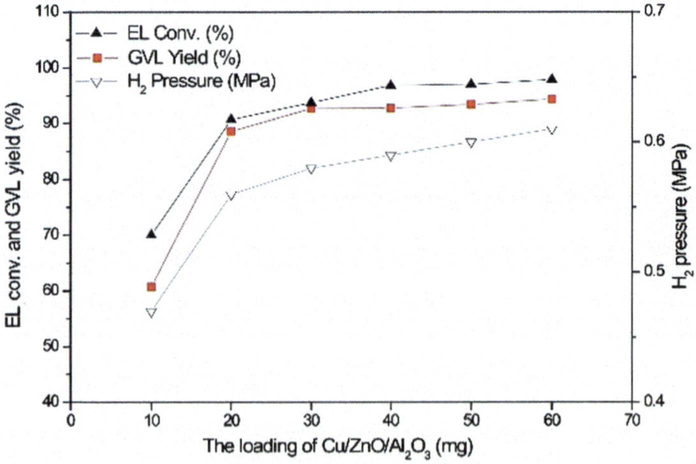

Fig. 2.11 The effect of CZA-3 loading (Conditions: 0.2 mmol EL, 2 mL *i*-PrOH, 120 °C, 3 h). H_2 pressure was calculated through the state equation of ideal gas (PV = nRT, T = 298 K) by the obtained concentration of acetone. Reprinted from Elsevier (Zhang et al., 2019)

0.5, 1, 2, 3 and 4 h (see Fig. 2.12a). At 140 °C, EL could be converted rapidly from 0.5 to 1 h, and EL conversion of 96.7% and GVL yield of 96.0% were achieved when time was 1 h. The optimal condition was 140 °C for 2 h to achieve quantitative conversion of EL and 99.0% yield of GVL. In addition, it could be seen that increasing reaction temperature or time could increase H_2 production and the GVL yield obviously (see Figs. 2.12 and 2.13). From the results above, it seemed to be a positive correlation between the amount of hydrogen and the yield of GVL. Considering the temperature tolerance of high pressure glass containers, the experiment at higher than 140 °C was not conducted.

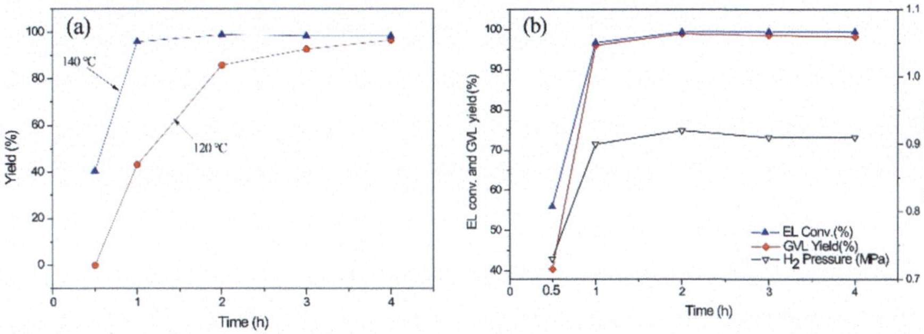

Fig. 2.12 **a** Effect of reaction time and temperature on EL conversion and GVL synthesis (Conditions: 0.2 mmol EL, 2 mL *i*-PrOH, 30 mg CZA-3); **b** Effect of reaction time for EL conversion and GVL synthesis at 140 °C. H2 pressure was calculated through the state equation of ideal gas (PV = nRT, T = 298 K) by the obtained concentration of acetone. Reprinted from Elsevier (Zhang et al., 2019)

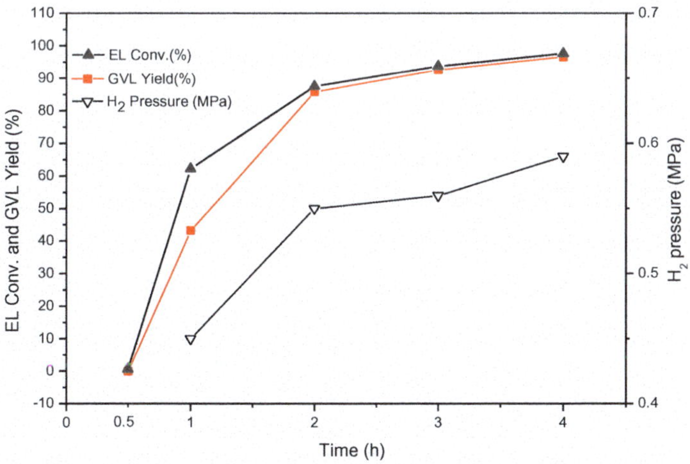

Fig. 2.13 Effect of reaction time for EL conversion and GVL synthesis at 120 °C. H_2 pressure was calculated through the state equation of ideal gas (PV = nRT, T = 298 K) by the obtained concentration of acetone. Reprinted from Elsevier (Zhang et al., 2019)

2.3.4 Reusability of the CZA-3 Catalyst

Since the lifetime of the catalyst had vital influence on economic benefit of whole reaction, the CZA-3 catalyst was recycled for several times to assess the reusability and stability. In Fig. 2.14, it could be observed that CZA-3 showed high reusability even after repeated four times without the loss of catalytic activity, and the yield kept almost unchanged in 99% with CZA-3 fresh and in 98% with CZA-3 after repeated four times. Furthmore, ICP analysis of the solution after the reaction illustrated that the stripping phenomenon of CZA-3 was negligible in the light of the concentration of Cu, Zn and Al were too small to be detected (Table 2.4, entry 1).

To check the stability of the catalyst, CZA-3 recovered after each CTH reaction were characterized by XRD, and the 4th recovered CZA-3 was characterized by BET, SEM, EDX and TEM.

The comparative study of the SEM morphology between CZA-3 (see Fig. 2.2b, c and d) and CZA-3 after reused four times (CZA-3R) (see Fig. 2.15a, b) illustrated that CZA-3R seemed to be rather similar with CZA-3. On the other hand, the elements distribution of Cu, Zn, Al and O were still uniform (see Fig. 2.15d). Crystallite size of Cu calculated by XRD patterns just had little change (12.7, 12.2, 12.6, 13.7 and

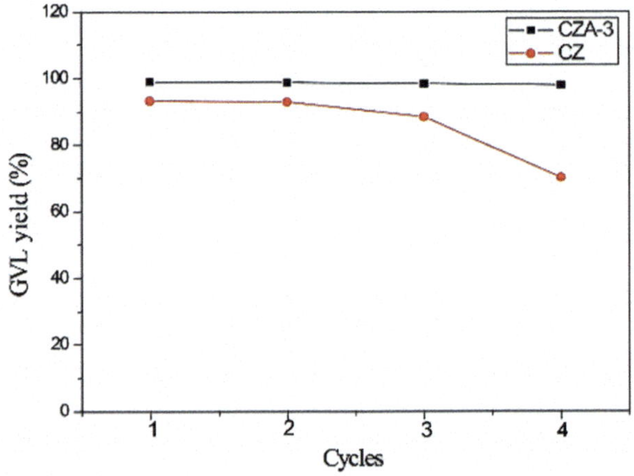

Fig. 2.14 Reuse stability of the CZA-3 and CZ catalysts (Conditions: 0.2 mmol EL, 40 mg catalysts, 2 mL *i*-PrOH, 140 °C, 5 h). Reprinted from Elsevier (Zhang et al., 2019)

Table 2.4 ICP analysis results

Entry	Cu (ppm)	Zn (ppm)	Al (ppm)
1[a]	0.1698	1.920	2.057
2[b]	17.79	195.6	48.45

Reprinted from Elsevier (Zhang et al., 2019)

[a] Liquid samples catalyzed by CZA-3, reaction conditions: 0.2 mmol EL, 30 mg CZA-3, 2 mL *i*-PrOH, 140 °C, 2 h

[b] Liquid samples catalyzed by CZA-3, reaction conditions: 0.2 mmol LA, 30 mg CZA-3, 2 mL *i*-PrOH, 140 °C, 2 h

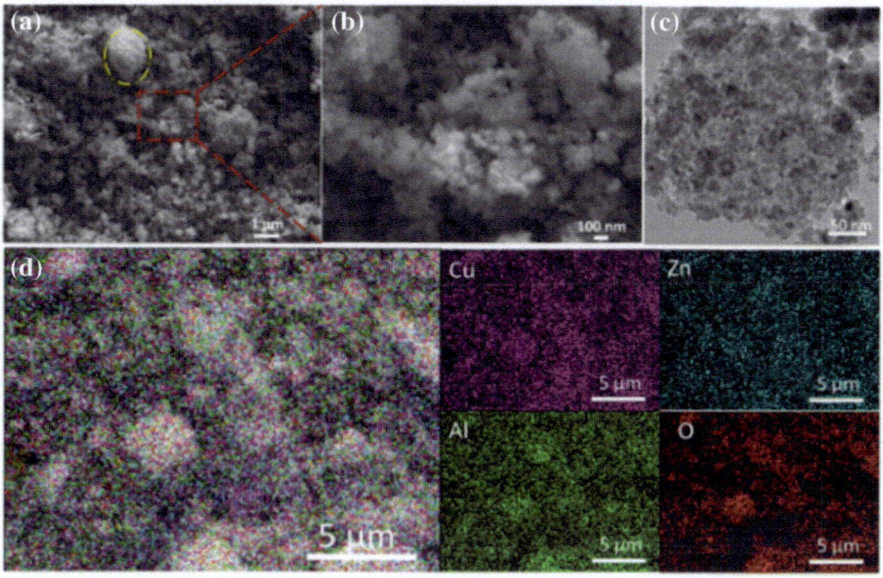

Fig. 2.15 **a**, **b** SEM images of CZA-3R; **c** TEM images and **d** EDX mapping of CZA-3R. Reprinted from Elsevier (Zhang et al., 2019)

14.0 nm) (see Fig. 2.16a), which did not have much influence on catalytic performance. Besides, CZA-3R had the superior surface area of 60.4 m^2/g compared to CZA-3 of 56.4 m^2/g, indicating that there was no collapse of porous structure, which might contribute to maintain the high catalytic activity of the CZA-3R (see Table 2.5). According to the characterization results above, the surface morphology and internal structure of CZA-3R did not change a lot in comparison to the fresh one. Thus, it was proved that CZA-3 had good stability and sustainability.

To explore the role of Al_2O_3 in the CZA-3 catalysts, the reusability of CZ catalysts was also tested in the same reaction conditions. As shown in Fig. 2.14, the activity of CZ catalysts decreased on subsequent uses. In Fig. 2.16b, the XRD pattern of the CZ catalyst after the use showed that the size of Cu particles grew from 15.1 to 17.9 nm and the size of ZnO particles grew from 9.4 to 12.5 nm. It seemed that the sintering of Cu after reused several times at high temperature might be one of the reasons of deactivation of catalyst (Kurtz et al., 2003). Since the phenomenon of Cu sintering of CZA-3 was obvious less than CZ, it indicated that Al_2O_3 had improved the thermal stability and sustainability of CZA-3 to maintain the catalytic activity, which was consistent with the previous studies (Guo et al., 2009; Shishido et al., 2006).

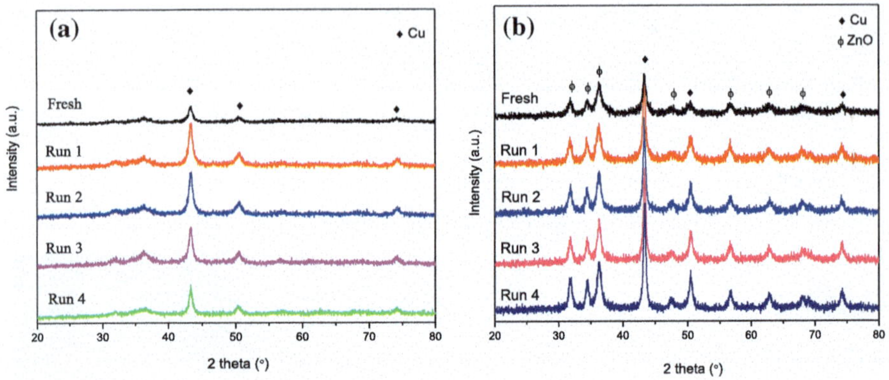

Fig. 2.16 XRD patterns of catalysts fresh and reused. **a** CZA-3 catalysts; **b** CZ catalysts. Reprinted from Elsevier (Zhang et al., 2019)

Table 2.5 Structural properties for CZA-3 and CZA-3R

Catalyst	N_2 adsorption/desorption			X-ray diffraction
	S_{BET}[a] (m^2/g)	V_p[b] (cc/g)	d_p[c] (nm)	d_{Cu}[d] (nm)
CZA-3	56.409	0.454	3.408	12.7
CZA-3R	60.424	0.39	3.056	14.0

Reprinted from Elsevier (Zhang et al., 2019)
[a] BET surface area. [b] Total pore volume determined as the N_2 adsorption. [c] Average pore diameter. [d] Cu mean crystallite size as determined by X-ray diffraction

2.3.5 Scopes of H-Donors and Raw Materials

There was no doubt that H-donors were very important for the CTH reaction. Among various primary and secondary alcohols estimated, only the secondary alcohols such as *i*-PrOH and 2-BuOH exhibited better performance, and similar yields of 99.0 and 97.2% were obtained, respectively (see Table 2.6, entries 1–2). On the other hand, primary alcohols were observed negative performances for the CTH reaction (entries 3–6). Moreover, Methyl levulinate and propyl levulinate were detected in the solution by GC-MS with MeOH or EtOH as H-donor, respectively, indicating that the transesterification reaction occurred instead of hydrocyclization of EL. What was interesting was that only cyclohexanol, EL, cyclohexanone and a little amount of GVL were detected when cyclohexanol was used as H-donor, indicating that EL was converted into the cyclohexanone (entry 6).

The feedstocks such as LA and its ester ML were also carried out (see Table 2.6, entries 7 and 8). The results showed that CZA-3 had a good performance for the CTH reaction of ML to obtain 97% yield of GVL. When LA was used as a feedstock, it was ineffective to transform to GVL and isopropyl levulinate was generated, indicating that the transesterification reaction occurred rather than CTH reaction. The ICP

Table 2.6 Various H-donors and feedstocks for the CTH reaction. Reprinted from Elsevier (Zhang et al., 2019)

Entry	Feedstock	H-donor	EL Conv. (%)	Yield (%)	
				GVL	PDO
1	EL	*i*-PrOH	99.4	99.0	0.4
2	EL	2-BuOH	98.8	97.2	–
3	EL	MeOH	37.2	2.5	–
4	EL	EtOH	3.7	3.4	–
5	EL	1-PrOH	18.4	3.9	–
6	EL	Cyclohexanol	73.0	2.7	–
7	ML	*i*-PrOH	100	97.0	–
8[b]	LA	*i*-PrOH	2.5	0.9	–

Conditions: 0.2 mmol feedstock, 30 mg CZA-3, 2 mL H-donor, 140 °C, 2 h

analysis of the solution illustrated that the concentrations of Cu, Zn and Al were 17.79, 195.6 and 48.45 ppm, respectively, it indicated that the support of CZA-3 was destroyed by LA which was responsible for the trace yield of GVL (see Table 2.4, entry 2).

2.3.6 The Proposed Mechanism of CTH Reaction of EL Over CZA-3

To figure out the mechanism of this CTH reaction, the liquid phase after the reaction was measured by GC chromatogram, and acetone as the product of dehydrogenation of *i*-PrOH and ethyl 4-hydroxypentanoate were detected in Fig. 2.17a. In addition, the gas phase was also collected after the reaction, an obvious H2 peak was detected by GC-TCD as shown in Fig. 2.17b. It was obvious that dehydrogenation of *i*-PrOH occurred to produce hydrogen and acetone (Al-Shaal et al., 2016; Kuwahara et al., 2014b; Zhen et al., 2013), the results verified that the pathway of CTH was the metal hydride route which was different from direct hydrogen transfer (Meerwein-Ponndorf-Verley, MPV mechanism) (Matthew et al., 2016). The obtained concentration of acetone was used to calculate H2 pressure through the state equation of ideal gas ($PV = nRT$, $T = 298$ K).

According to our results and previous studies (Gilkey & Xu, 2016; Xue et al., 2016; Zhen et al., 2013), the pathway for the CTH reaction of EL over CZA-3 was proposed as shown in Fig. 2.18. Firstly, a surface alkoxy group was generated when the O–H bond of *i*-PrOH was absorbed on CZA-3 surface. Then, β-H elimination happened to produce acetone and two Cu-H active species of CTH process formed. The active H species hydrogenated acetyl group to obtain ethyl 4-hydroxypentanoate (D). Finally, intramolecular nucleophilic attack of compound **D** took place to afford

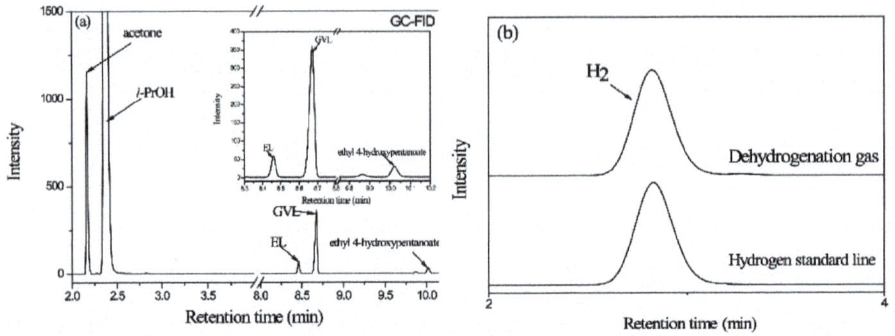

Fig. 2.17 **a** GC-FID of liquid samples catalyzed by CZA-3, reaction condition: 0.2 mmol EL, 20 mg catalyst, 2 mL *i*-PrOH, 120 °C, 3 h; **b** GC-TCD of dehydrogenation gas catalyzed by CZA-3. Reprinted from Elsevier (Zhang et al., 2019)

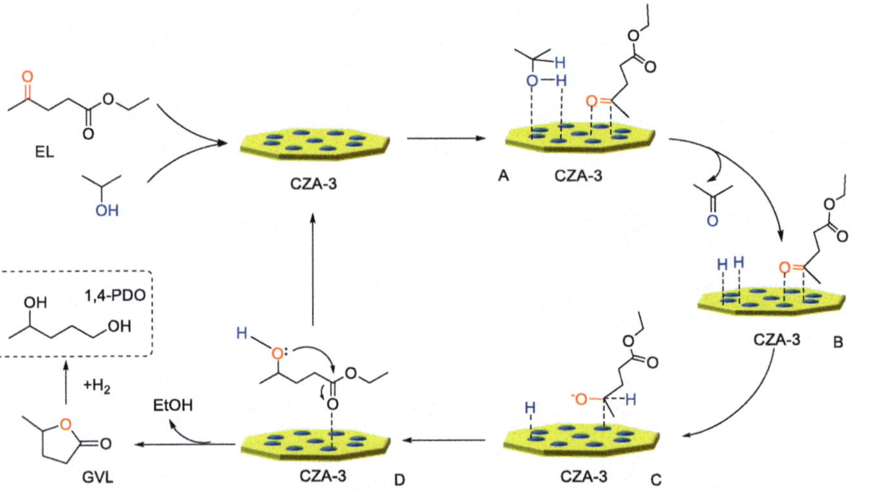

Fig. 2.18 The proposed pathway for the hydrogenation of EL to GVL over CZA-3. Reprinted from Elsevier (Zhang et al., 2019)

GVL along with the regeneration of CZA-3 and the formation of ethanol. A trace amount of 1,4-PDO was also obtained by further hydrogenation of GVL (Ren et al., 2016).

2.4 Conclusions

We developed an effective ternary $Cu/ZnO/Al_2O_3$ catalyst by two-step co-precipitation method for CTH process of EL to GVL.

Comparing the various characterization results of CZA-3, CZ and CA, it seemed that ZnO and Al_2O_3 as binary support could effectively regulate the specific surface area and the crystallite size of Cu of the catalysts. CZA-3 catalyst with the good dispersion of small Cu metal nanoparticles and relatively large BET surface area not only had the highest catalytic activity for the CTH process, but also had good circulation stability that could be reused at least four times without the loss of catalytic activity. In addition, the catalytic activities of CZA catalysts in CTH process of EL were closely related to the precursor calcination temperature which determined the crystallite size of Cu and its porous structure. Experimental results exhibited that lower precursor calcination temperature was needed for the reaction. With optimized conditions in hand, CZA-3 gave a 99.0% yield of GVL with the *i*-PrOH as an H-donor at 140 °C for 2 h. The present study provides a low cost, efficient and environmentally nanoparticle catalyst for the dehydrogenation process of *i*-PrOH and hydrogenation process of EL.

References

Al-Shaal, M. G., Calin, M., Delidovich, I., & Palkovits, R. (2016). Microwave-assisted reduction of levulinic acid with alcohols producing γ-valerolactone in the presence of a Ru/C catalyst. *Catalysis Communications, 75*, 65–68.

Amarasekara, A. S., & Hasan, M. A. (2015). Pd/C catalyzed conversion of levulinic acid to γ-valerolactone using alcohol as a hydrogen donor under microwave conditions. *Catalysis Communications, 60*, 5–7.

Atake, I., Nishida, K., Li, D., Shishido, T., Oumi, Y., Sano, T., & Takehira, K. (2007). Catalytic behavior of ternary Cu/ZnO/Al_2O_3 systems prepared by homogeneous precipitation in water-gas shift reaction. *Journal of Molecular Catalysis a: Chemical, 275*, 130–138.

Bahmani, M., Vasheghani Farahani, B., & Sahebdelfar, S. (2016). Preparation of high performance nano-sized Cu/ZnO/Al_2O_3 methanol synthesis catalyst via aluminum hydrous oxide sol. *Applied Catalysis A: General, 520***,** 178–187.

Baltes, C., Vukojevic, S., & Schuth, F. (2008). Correlations between synthesis, precursor, and catalyst structure and activity of a large set of CuO/ZnO/Al_2O_3 catalysts for methanol synthesis. *Journal of Catalysis, 258*, 334–344.

Chia, M., & Dumesic, J. A. (2011). Liquid-phase catalytic transfer hydrogenation and cyclization of levulinic acid and its esters to gamma-valerolactone over metal oxide catalysts. *Chemical Communications, 47*, 12233–12235.

De, S., Saha, B., & Luque, R. (2015). Hydrodeoxygenation processes: Advances on catalytic transformations of biomass-derived platform chemicals into hydrocarbon fuels. *Bioresource Technology, 178*, 108–118.

Fichtl, M. B., Schlereth, D., Jacobsen, N., Kasatkin, I., & Hinrichsen, O. (2015). Kinetics of deactivation on Cu/ZnO/Al_2O_3 methanol synthesis catalysts. *Applied Catalysis a: General, 502*, 262–270.

Figueiredo, R. T., Martínez-Arias, A., Granados, M. L., & Fierro, J. (1998). Spectroscopic evidence of Cu–Al interactions in Cu–Zn–Al mixed oxide catalysts used in CO hydrogenation. *Journal of Catalysis, 178*, 146–152.

Gilkey, M. J., & Xu, B. (2016). Heterogeneous catalytic transfer hydrogenation as an effective pathway in biomass upgrading. *ACS Catalysis, 6*, 1420–1436.

Guo, P. J., Chen, L. F., Yu, G. B., Yuan, Z., Qiao, M. H., Xu, H. L., & Fan, K. N. (2009). Cu/ZnO-based water–gas shift catalysts in shut-down/start-up operation. *Catalysis Communications, 10*, 1252–1256.

Hengne, A. M., Kadu, B. S., Biradar, N. S., Chikate, R. C., & Rode, C. V. (2016). Transfer hydrogenation of biomass-derived levulinic acid to γ-valerolactone over supported Ni catalysts. *RSC Advances, 6*, 59753–59761.

Hu, L., Lin, L., Wu, Z., Zhou, S., & Liu, S. (2017). Recent advances in catalytic transformation of biomass-derived 5-hydroxymethylfurfural into the innovative fuels and chemicals. *Renewable and Sustainable Energy Reviews, 74*, 230–257.

Igor, K., Patrick, K., Benjamin, K., Annette, T., & Schlögl, R. (2007). Role of lattice strain and defects in copper particles on the activity of $Cu/ZnO/Al_2O_3$ catalysts for methanol synthesis. *Angewandte Chemie, 46*, 7324–7327.

Khan, A. S., Man, Z., Bustam, M. A., Kait, C. F., Nasrullah, A., Ullah, Z., Sarwono, A., Ahamd, P., & Muhammad, N. (2018). Dicationic ionic liquids as sustainable approach for direct conversion of cellulose to levulinic acid. *Journal of Cleaner Production, 170*, 591–600.

Kowalik, P., & Próchniak, W. (2010). The effect of calcination temperature on properties and activity of $Cu/ZnO/Al_2O_3$ catalysts. *Annales UMCS, Chemistry, 65*, 79–87.

Kurtz, M., Wilmer, H., Genger, T., Hinrichsen, O., & Muhler, M. (2003). Deactivation of supported copper catalysts for methanol synthesis. *Catalysis Letters, 86*, 77–80.

Kuwahara, Y., Kaburagi, W., & Fujitani, T. (2014a). Catalytic conversion of Levulinic acid and its esters to γ-valerolactone over silica-supported zirconia catalysts. *Bulletin of Chemical Society of Japan, 87*, 1252–1254.

Kuwahara, Y., Kaburagi, W., & Fujitani, T. (2014b). Catalytic transfer hydrogenation of levulinate esters to γ-valerolactone over supported ruthenium hydroxide catalysts. *RSC Advances, 4*, 45848–45855.

Kuwahara, Y., Kaburagi, W., Osada, Y., Fujitani, T., & Yamashita, H. (2017a). Catalytic transfer hydrogenation of biomass-derived levulinic acid and its esters to γ-valerolactone over ZrO_2 catalyst supported on SBA-15 silica. *Catalysis Today, 281*, 418–428.

Kuwahara, Y., Kango, H., & Yamashita, H. (2017b). Catalytic transfer hydrogenation of biomass-derived Levulinic acid and its esters to γ-valerolactone over sulfonic acid-functionalized UiO-66. *ACS Sustainable Chemistry & Engineering, 5*, 1141–1152.

Li, H., Fang, Z., Jian, H., & Yang, S. (2017). Orderly layered Zr-benzylphosphonate nanohybrids for efficient acid–base-mediated bifunctional/cascade catalysis. *Chemsuschem, 10*, 681–686.

Li, Z., Tang, X., Jiang, Y., Wang, Y., Zuo, M., Chen, W., Zeng, X., Sun, Y., & Lin, L. (2015). Atom-economical synthesis of gamma-valerolactone with self-supplied hydrogen from methanol. *Chemical Communications, 51*, 16320–16323.

López-Suárez, F. E., Bueno-López, A., & Illán-Gómez, M. J. (2008). Cu/Al_2O_3 catalysts for soot oxidation: Copper loading effect. *Applied Catalysis B: Environmental, 84***,** 651–658.

Luo, H. Y., Consoli, D. F., Gunther, W. R., & Román-Leshkov, Y. (2014). Investigation of the reaction kinetics of isolated Lewis acid sites in Beta zeolites for the Meerwein–Ponndorf–Verley reduction of methyl levulinate to γ-valerolactone. *Journal of Catalysis, 320*, 198–207.

Matthew, J., Gilkey, B., & Xu, B. (2016). Heterogeneous catalytic transfer hydrogenation as an effective pathway in biomass upgrading. *ACS Catalysis, 6*, 1420–1436.

Ren, D., Wan, X., Jin, F., Song, Z., Liu, Y., & Huo, Z. (2016). Selective hydrogenation of levulinate esters to 1,4-pentanediol using a ternary skeletal CuAlZn catalyst. *Green Chemistry, 18*, 5999–6003.

Ren, H., Xu, C., Zhao, H., Wang, Y., Liu, J., & Liu, J. (2015). Methanol synthesis from CO_2 hydrogenation over $Cu/\gamma\text{-}Al_2O_3$ catalysts modified by ZnO, ZrO_2 and MgO. *Journal of Industrial and Engineering Chemistry, 28*, 261–267.

Shishido, T., Yamamoto, M., Li, D., Tian, Y., Morioka, H., Honda, M., Sano, T., & Takehira, K. (2006). Water-gas shift reaction over Cu/ZnO and $Cu/ZnO/Al_2O_3$ catalysts prepared by homogeneous precipitation. *Applied Catalysis a: General, 303*, 62–71.

Song, J., Wu, L., Zhou, B., Zhou, H., Fan, H., Yang, Y., Meng, Q., & Han, B. (2015). A new porous Zr-containing catalyst with a phenate group: An efficient catalyst for the catalytic transfer hydrogenation of ethyl levulinate to g-valerolactone. *Green Chemistry, 17*, 1626–1632.

Tang, X., Chen, H., Hu, L., Hao, W., Sun, Y., Zeng, X., Lin, L., & Liu, S. (2014). Conversion of biomass to γ-valerolactone by catalytic transfer hydrogenation of ethyl levulinate over metal hydroxides. *Applied Catalysis b: Environmental, 147*, 827–834.

Tang, X., Li, Z., Zeng, X., Jiang, Y., Liu, S., Lei, T., Sun, Y., & Lin, L. (2015). In situ catalytic hydrogenation of biomass-derived methyl levulinate to gamma-valerolactone in methanol. *Chemsuschem, 8*, 1601–1607.

Tang, X., Wei, J., Ding, N., Sun, Y., Zeng, X., Hu, L., Liu, S., Lei, T., & Lin, L. (2017). Chemoselective hydrogenation of biomass derived 5-hydroxymethylfurfural to diols: Key intermediates for sustainable chemicals, materials and fuels. *Renewable and Sustainable Energy Reviews, 77*, 287–296.

Wang, Y., Rong, Z., Wang, Y., Wang, T., Du, Q., Wang, Y., & Qu, J. (2016). Graphene-based metal/acid bifunctional catalyst for the conversion of Levulinic acid to γ-valerolactone. *ACS Sustainable Chemistry & Engineering, 5*, 1538–1548.

Xue, Z., Jiang, J., Li, G., Zhao, W., Wang, J., & Mu, T. (2016). Zirconium–cyanuric acid coordination polymer: Highly efficient catalyst for conversion of levulinic acid to γ-valerolactone. *Catalysis Science & Technology, 6*, 5374–5379.

Yang, Y., Xu, X., Zou, W., Yue, H., Tian, G., & Feng, S. (2016). Transfer hydrogenation of methyl levulinate into gamma-valerolactone, 1,4-pentanediol, and 1-pentanol over Cu–ZrO_2 catalyst under solvothermal conditions. *Catalysis Communications, 76*, 50–53.

Yuan, J., Li, S., Yu, L., Liu, Y., Cao, Y., He, H., & Fan, K. (2013). Copper-based catalysts for the efficient conversion of carbohydrate biomass into γ-valerolactone in the absence of externally added hydrogen. *Energy & Environmental Science, 6*, 3308–3313.

Zhang, Q., Wu, T., Zhang, P., Qi, R., Huang, R., Song, X., & Gao, L. (2014). Facile synthesis of hollow hierarchical Ni/γ-Al_2O_3 nanocomposites for methane dry reforming catalysis. *RSC Advances, 4*, 51184–51193.

Zhang, Y., Chen, C., Gong, W., Song, J., Zhang, H., Zhang, Y., Wang, G., & Zhao, H. (2017). Self-assembled Pd/CeO_2 catalysts by a facile redox approach for high-efficiency hydrogenation of levulinic acid into gamma-valerolactone. *Catalysis Communications, 93*, 10–14.

Zhang, C., Huo, Z., Ren, D., Song, Z., Liu, Y., Jin, F., & Zhou, W. (2019). Catalytic transfer hydrogenation of levulinate ester into γ-valerolactone over ternary Cu/ZnO/Al_2O_3 catalyst. *Journal of Energy Chemistry, 32*, 189–197.

Zhen, Y., Huang, Y. B., Guo, Q. X., & Yao, F. (2013). RANEY® Ni catalyzed transfer hydrogenation of levulinate esters to γ-valerolactone at room temperature. *Chemical Communications, 49*, 5328–5330.

Chapter 3
Catalytic Transfer Hydrogenation of Ethyl Levulinate into γ-Valerolactone Over Air-Stable Skeletal Cobalt Catalyst

3.1 Introduction

Plant-based biomass is a sustainable and low-carbon resources to produce fuel and chemicals (Cao et al., 2020). Levulinic acid (LA) and its esters are versatile biomass platform molecules, can be synthesized from cheap and abundant plant-based carbohydrate via simple acid catalysis (Pileidis & Titirici, 2016). Production of γ-valerolactone (GVL) by a LA hydrogenation process is one of the most important usage of LA, which is a hot area in catalytic biomass conversion in the past ten years (Sun et al., 2018; Qi & Horváth, 2012). Because GVL is a valuable organic compound and can be used as precursor for value-added chemicals and fuels production (Horváth et al., 2008). Although H_2 as H-donor has been widely used in upgrading bio-derived compounds, the potential safety issues in storage and transportation cannot be neglected (Zhang et al., 2018). Recently, many research groups focused on catalytic transfer hydrogenation (CTH) of biomass compounds to synthesize GVL using secondary alcohols such as 2-butanol (2-BuOH) and *i*-propanol (*i*-PrOH) to replace H_2 as H-donors. Compared with other H-donors (formic acid and methanol)(Gou et al., 2020), secondary alcohols are noncorrosive, low toxic, and no greenhouse gas (CO_2) release during the reaction, the dehydrogenation products ketone can be reused in chemical industry.

Transition metal-based catalysts are popular in hydrogenation of levulinate esters with high pressure gaseous H_2, the processes achieved, however, by CTH route are relatively less. Among these studies, supported Ru (Huang et al., 2018), Pd (Amarasekara & Hasan, 2015), Ni (Hengne et al., 2016), and Cu (Yang et al., 2016) catalysts were investigated, but Ru and Pd are costly, and cheap Ni and Cu catalysts should be activated by reductant before it is used. Hence, Fu et al. firstly reported an non-precious skeletal Ni catalyst which could effective catalyze the reaction with *i*-PrOH as H-donor at room temperature over 9 h (Yang et al., 2013), the catalyst activation process is not necessary before the reaction set up. However, skeletal metals are air sensitive and prone to spontaneous combustion when it is exposed in air. Therefore, development of air-stable skeletal metal catalysts is highly desirable

Z. Huo, *Diverse Hydrogen Sources for Biomass-derivatives Conversion*,
https://doi.org/10.1007/978-981-99-1673-3_3

for conventional biomass hydrogenation or CTH process. Passivation techniques is commonly applied to preserve air sensitive transition metal catalysts, avoid violent catalyst oxidation, and maintain the catalyst activity (Hammache et al., 2002). Metal passivation process is often conducted in inert gas flow with low concentration of O_2 (0.1–1%) or other mildly oxidizing atmosphere such as CO_2, N_2O and H_2O to achieve air stable catalysts (Wolf et al., 2016). However, tube furnace with good gas tightness and precise controlling the concentration of mixed gas are essential for valid passivation treatment, which increases the cost and complexity (Huber et al., 2006). Furthermore, the mildly oxidizing gas flow process is time-consuming, which usually takes several hours or days. Therefore, it is highly recommended for developing fast and low-cost passivation treatment for air sensitive metal catalyst. Recently, Co-based catalysts were reported for CTH of carbonyl compounds (Kumar et al., 2018; Suresh Kumar et al., 2017), N-heteroarenes (Chen et al., 2017), nitrobenzene (Guo et al., 2019; Yuan et al., 2018), unsaturated fatty acids (Wang et al., 2019) with formic acid or *i*-PrOH. Therefore, non-precious skeletal Co (SCo) is an attractive candidate for levulinate esters CTH process.

Herein, a batch air diffusion method (BA) was developed to treat commercial SCo to prepare an air-stable skeletal Co (A-SCo). The BA process is fast and easy handled without special furnace and longtime inert gas flow treatment. The as-prepared A-SCo preserved good morphology structure and surface chemical state, which exhibited excellent performance for CTH of ethyl levulinate (EL) into GVL using H-ZSM-5 and i-PrOH as cocatalyst and H-donor, respectively. The BA process doesn't reduce the activity of SCo and no deactivation was obviously observed for A-SCo which was prepared and exposed in air atmosphere for one month.

3.2 Experimental

3.2.1 Materials

The commercial skeletal Co (SCo, 93 wt% Co, 7 wt% Al) and CoAl alloy were purchased from Dalian Tongyong Chemical Co., Ltd. Ethyl levulinate (98%) and AR grade of Co, Ni, Al, Co_2O_3, CoO, NiO were purchased from Sinopharm Chemical Reagent (China). HZSM-5 was purchased from Nankai University Catalyst Co., Ltd.

3.2.2 Pretreated Process for SCo

Preparation of A-SCo (batch air diffusion method (BA)). About 200 mg wet SCo was washed by ethanol and dried with nitrogen flow in a 5 mL centrifuge tube. Then, the tube was opened to allow air diffuse naturally into the tube for 5 s, the tube was sealed and rotated for 30 s. Finally, the same process should be repeated 4 times to

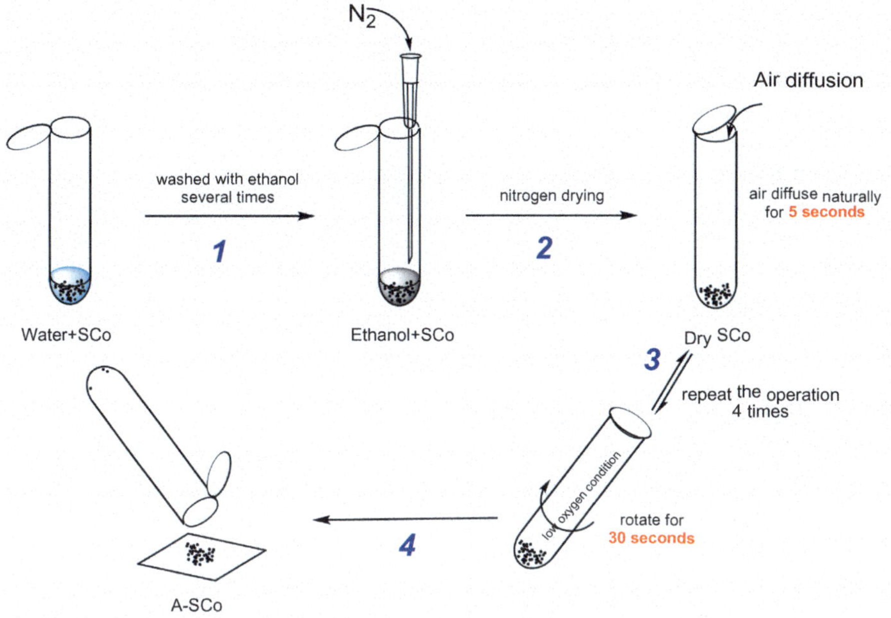

Fig. 3.1 The air pretreatment process of SCo. Reprinted from Elsevier (Ren, 2022)

obtain an A-SCo. The whole air-pretreated process was less than 5 min as shown in Fig. 3.1.

Preparation of vacuum-treated dry SCo. About 200 mg wet SCo was washed by ethanol in a centrifuge tube and loaded the tube into a nitrogen purged vacuum drying oven. And dried it at room temperature for 24 h. This vacuum-treated SCo was named as V-SCo.

Preparation of spontaneous combustion SCo. About 200 mg wet SCo was washed by ethanol in a centrifuge tube, then the dry SCo was exposed into air directly. Then, spontaneous combustion happened. After the ignited-SCo cooling down to room temperature, the spontaneous combustion SCo was obtained and named as S-SCo.

3.2.3 Catalyst Characterization

The catalyst phase identification was measured via X-ray diffraction (XRD) test. The as-prepared powder of SCo was placed into a sample pool and analyzed by LabX XRD-6100 (Shimadzu) using a CuKα radiation source. The samples were scanned over a 2θ range of 10–80° with the scanning speed of 1o min-1. Particularly, in order to avoid spontaneous combustion of Fresh-SCo, the sample need to stay moist for XRD test. The specific surface area and pore distribution of catalyst was measured by N_2 adsorption–desorption isotherms using an automated gas sorption analyzer (Autosorb-iQ-TPX, Quantachrome, USA) with BET (Brunauer–Emmett–Teller) method. Before testing, the samples were first pretreated with a degassing process under vacuum in a sample tube. The morphology of as-prepared catalysts was characterized by Transmission electron microscopy (TEM, JEOL JEM-2010). The samples were milled and dispersed homogenously in ethanol by ultrasonic before TEM test. The surface chemical state of catalysts was investigated via X-ray photoelectron spectroscopy (XPS, AXIS UltraDLD). The samples were also milled and fixed onto specialized adhesive tape for XPS test. CasaXPS and Jade were used to process the data of XPS and XRD, respectively.

3.2.4 Experimental Procedures

The CTH of EL experiments were conducted into a 6 mL high pressure glass reactor (Fig. 3.2). After feedstock (EL, 0.2 mmol), desired amount of catalyst and i-PrOH (2 mL, 26.2 mmol) were loaded, the reactor was purged with N_2 and sealed, then put it onto the preheated heater with magnetic stirring (Fig. 3.1). After the reaction, the reactor was cooled to room temperature. The liquid sample was collected, filtered with a 0.45 μm membrane and analyzed by GC-FID (gas chromatography with flame ionization detector, Agilent GC7890A), and GC–MS (gas chromatography with mass spectrometer, Agilent GC7890A- MS5975C) which equipped with HP-Innowax column (30 m × 0.25 mm × 0.25 μm). The GC operating parameter settings were as follows: 1 μL sample was injected into the injector port which was set at 260 °C, and the initial column temperature was set to 70 °C for 1 min, then raised to 220 °C with heat rate of 10 °C and maintained 220 °C for another 5 min. The solid catalysts were recovered and washed with water and ethanol for several times, then kept into ethanol. Procedure of i-PrOH dehydrogenation reaction is follow as the CTH of EL process. The details of reaction condition are listed below the relative Tables. Gaseous products were collected with syringe and analyzed by GC-TCD.

The definition of feedstock conversion and product yield is as follows:

$$\text{Conversion}_{\text{EL}}(\%) = \left(1 - \frac{\text{final amount of EL (mmol)}}{\text{initial amount of EL (mmol)}}\right) \times 100(\%)$$

$$\text{Conversion}_{i-\text{PrOH}}(\%) = \frac{\text{final amount of acetone (mmol)}}{\text{initial amount of } i - \text{PrOH (mmol)}} \times 100(\%)$$

$$\text{Yield}_{\text{GVL}}(\%) = \frac{\text{final amount of GVL (mmol)}}{\text{initial amount of EL (mmol)}} \times 100(\%)$$

H_2 pressure was calculated via the state of equation of ideal gas (PV = nRT). Because 1 mol H_2 and 1 mol acetone are produced via the dehydrogenation of 1 mol i-PrOH, the production of H_2 can be Figureured out by the content of acetone. Hence, the pressure of H_2 can be calculated by the state of equation of

Fig. 3.2 The glass reactor of CTH of EL. Reprinted from Elsevier (Ren, 2022)

ideal gas (PV = nRT). For instance, 1 mmol acetone is quantitated, the volume of the glass reactor and i-PrOH are 6 mL and 2 mL, respectively. The generated H_2 pressure at room temperature (25 °C) is calculated:

$$P_{H2} = \frac{nRT}{V} = \frac{1(\text{mmol}) \times 10^{-3} \times 8.31\left(\text{Pa} * \text{m}^3 * \text{mol}^{-1} * \text{K}^{-1}\right) \times (273 + 25)(\text{K})}{(6-2)(\text{mL}) \times 10^{-6}} \times 10^{-6} = 0.62(\text{MPa})$$

3.3 Results and Discussion

3.3.1 *Characterizations of Fresh-SCo, V-SCo, A-SCo and S-SCo*

Spontaneous combustion is common property of Fresh-SCo which was confirmed by air exposure test (Fig. 3.3a, Movie S1). After pretreated with batch air diffusion process (BA), A-SCo exhibited high air stability (Fig. 3.3b, Movie S2). It should be noted that after 24 h vacuum drying, V-SCo is also stable in air condition. In order to understand the effects of different pretreatments on Fresh-SCo. XRD tests were first conducted and the intensity of diffraction peaks were normalized to display in Fig. 3.3c. Only 3 diffraction peaks of metallic Co (PDF#15–0806) are observed both in Fresh-SCo, V-SCo and A-SCo, no Co or Al oxides are displayed, indicating that metallic Co in bulk phase of A-SCo wasn't oxidized via BA process. However, the XRD pattern of S-SCo gives three new peaks apart from Co peaks which might be Co and Al oxides. As shown in Fig. 3.3d, N2 adsorption–desorption curves of V-SCo and A-SCo are similar which means the surface structure of SCo maintained after air pretreatment. Furthermore, the much lower adsorption volume of S-SCo can be attributed to the spontaneous combustion to complete destroy the skeletal structure. Specific surface areas of V-SCo, A-SCo and S-SCo are 27.7 m^2/g and 27.2 m^2/g, 10.4 m^2/g, respectively, indicating that the structure of SCo wasn't significantly altered by air pretreatment. As shown in Fig. 3.4, the mean pore diameter of V-SCo, A-SCo and S-SCo are 9.6, 6.5 and 9.6 nm, respectively, which indicates that BA process reduced the pore size of catalyst slightly. However, the peaks intensity of pore size distribution curves between V-SCo and A-SCo are close, demonstrating the numerous pores were preserved during the BA process. Moreover, low peak intensity of S-SCo indicates the number of pores decreases obviously, implying the structure of SCo was destroyed after pyrophoric process happened.

TEM tests were conducted to investigate the morphology of Fresh-SCo, A-SCo and S-SCo. Due to the air sensitive character of Fresh-SCo, V-SCo was used to replace Fresh-SCo for TEM analysis. The morphologies of V-SCo, A-SCo and S-SCo are displayed in Figure 3.5. V-SCo and A-SCo show flake shape with strip-like pores (marked with white circle). The S-SCo morphology in Fig. 3.5c shows that aggregation and fragmentation are observed, which is totally different from V-SCo

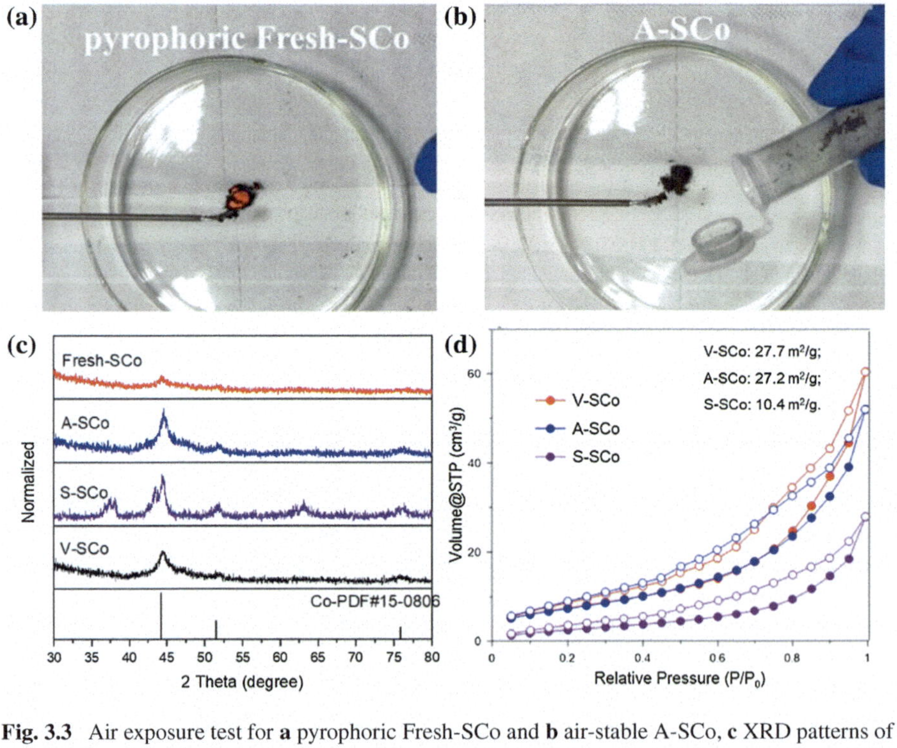

Fig. 3.3 Air exposure test for **a** pyrophoric Fresh-SCo and **b** air-stable A-SCo, **c** XRD patterns of skeletal Co with different pretreatment, **d** N_2 adsorption and desorption isotherms of Fresh-SCo, A-SCo and S-SCo. Reprinted from Elsevier (Ren, 2022)

Fig. 3.4 Distribution of pore size of V-SCo, A-SCo and S-SCo (calculated by BJH model). Reprinted from Elsevier (Ren, 2022)

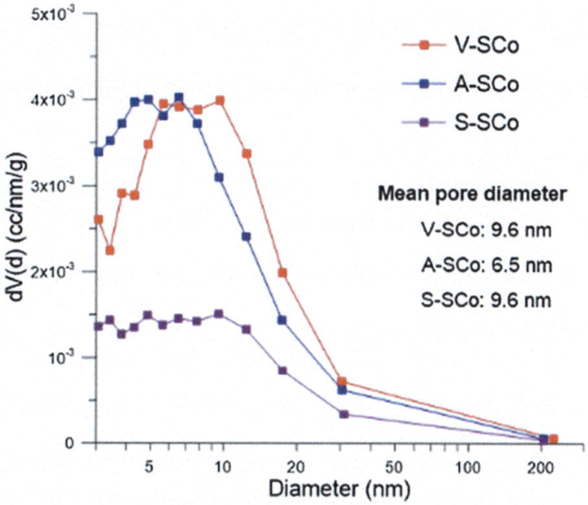

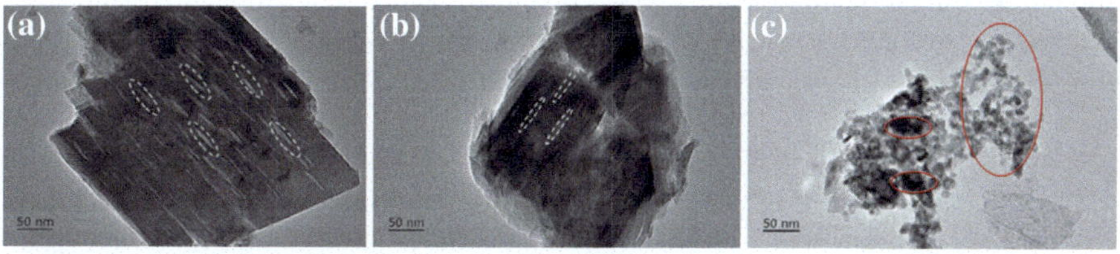

Fig. 3.5 TEM images of **a** V-SCo, **b** A-SCo and **c** S-SCo (nanopores are marked with white circles, aggregation and fragmentation are marked with red circles). Reprinted from Elsevier (Ren, 2022)

and A-SCo indicating that the structure of SCo is sintered and destroyed completely. This is entirely consistent with the results of N_2 adsorption and desorption isotherms.

Then XPS was also performed to examine the surface chemical state of SCo with different treatments. The same as TEM analysis, V-SCo was used instead of Fresh-SCo for XPS analysis. XPS survey profiles of V-SCo, A-SCo, and S-SCo are shown in Fig. 3.6. 3.7 mol%, 3.8 mol% and 2.9 mol% of Co species are detected over these pretreated catalysts, respectively. While Al species is the major part on the surface of these catalysts, the mole ratio of Al exceeds 50%, even 93 wt% Co and 7 wt% Al are recorded for SCo. In addition, small amount of Ni (about 1 mol%) is found on these commercial catalysts. High resolution XPS spectra of Co $2p_{3/2}$, Ni $2p_{3/2}$, Al 2p, O 1 s derived from V-SCo, A-SCo and S-SCo are displayed in Fig. 3.7. The chemical state of Co is a key factor for hydrogenation of EL (Zhou et al., 2014). The Co $2p_{3/2}$ spectrum of V-SCo can be deconvoluted into four peaks at 777.5, 778.9, 781.1 and 785.2 eV which are assigned to metallic Co, Co(III), Co(II) and shake up satellite peak, respectively (Fig. 3.7a) (Fierro et al., 2000; Riva et al., 2000). The ratio of metallic Co/Co(II/III) is 0.25. In comparison, metallic Co, oxidized Co species are observed obviously from the Co $2p_{3/2}$ spectrum of A-SCo, only a slightly decrease

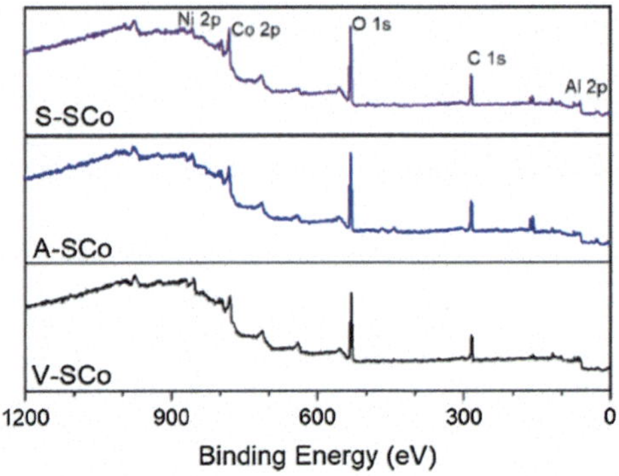

Fig. 3.6 The XPS survey of V-SCo, A-SCo and S-SCo. Reprinted from Elsevier (Ren, 2022)

of Co/Co(II/III) (0.22) is discovered, indicating that a plenty of metallic Co exists on the surface of A-SCo via BA process which ensures the active sites for ETH reaction. As expected, no metallic Co peak but a mass of Co(III) are found from Co $2p_{3/2}$ spectrum of S-SCo, demonstrating that metallic Co on the surface of SCo was complete oxidized during the self-ignite process. The same findings are discovered from the spectra of Ni $2p_{3/2}$, Al 2p. As is shown in Fig. 3.7b, three peaks of Ni $2p_{3/2}$ from V-SCo, A-SCo and S-SCo located around 852.4, 855.5 and 860.8 eV which are assigned to metallic Ni, Ni (II/III) and satellite peak (Ni et al., 2019). The term of Ni/Ni(II/III) are 0.34, 0.27 and 0.06, respectively. For Al 2p spectra showing in Fig. 3.7c, the appearance of peaks at 74.2 eV implies the existence of Al oxides on the SCo surface (Chen et al., 2000). Additionally, the peaks around 71.5 eV assigned to metallic Al (Feliu & Barranco, 2003) are only observed in V-SCo and A-SCo, which indicates that metallic Al can be maintained after air or vacuum pretreatment. Two peaks of O 1 s spectra are fitted for V-SCo, A-SCo and S-SCo in Fig. 3.7d, the peak located at 529.3 eV corresponds to metal oxides, and the other at around 531.6 eV assigned to carbon–oxygen bonds (Zhang et al., 2010). It is obvious that the self-ignited S-SCo contains more metal oxides than that of V-SCo, A-SCo. From the analysis above, BA process is appropriate method to avoid spontaneous combustion of SCo and maintain the morphology, structure and surface chemical state of SCo.

3.3.2 Catalysts Performance for the EL Conversion

A-SCo and other Co, Al based catalysts were selected to test the catalytic activity for GVL production, and the results were summarized in Table 3.1. The blank test was first conducted, no EL and i-PrOH were consumed, and no acetone, a dehydrogenation product of i-PrOH, was detected, which verifies that EL and i-PrOH cannot react with each other spontaneously (entry 1). After A-SCo was added, EL was totally converted and 68% yield of GVL was obtained after reaction (entry 2). From Fig. 3.8a, a mass of acetone and an unknown product were observed by GC chromatogram. This unknown product might be assigned to ethyl 4-hydroxypentanoate which was

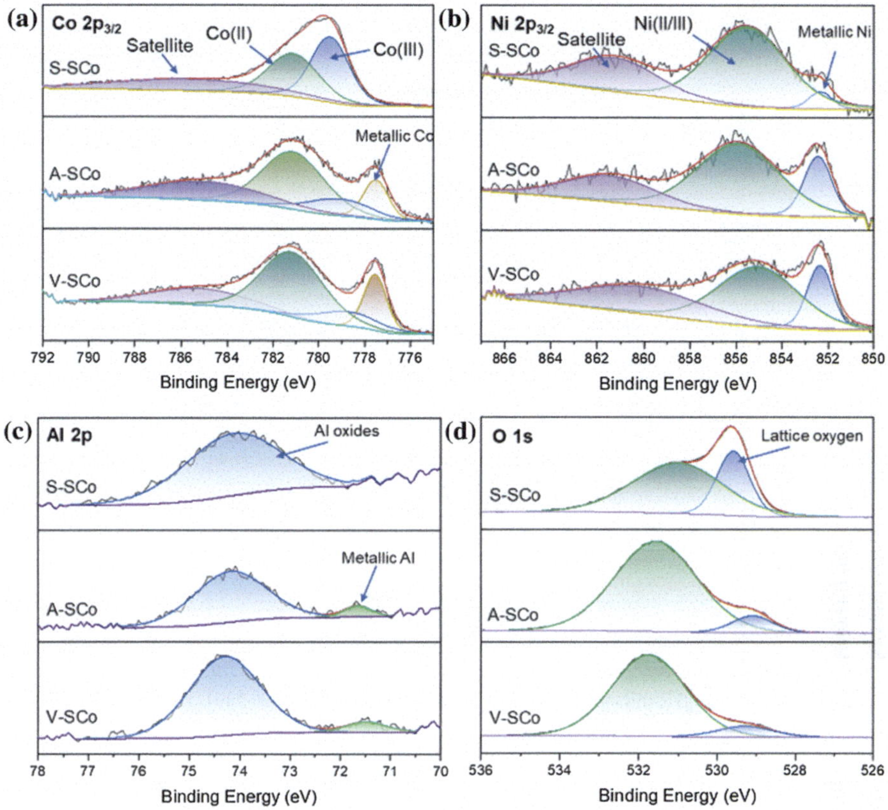

Fig. 3.7 **a** Co $2p_{3/2}$, **b** Ni $2p_{3/2}$, **c** Al 2p and **d** O 1 s XPS spectra for V-SCo, A-SCo and S-SCo. Reprinted from Elsevier (Ren, 2022)

considered as an intermediate (Xu et al., 2018). Because the mass spectrometry result (Fig. 3.9) cannot match the specific organic molecular from database, ion fragments information is another evidence for analyzing the structure of molecules. Ethyl 4-hydroxypentanoate is the promising candidate for this unknown product. However, the maximum m/z is 131 cannot be confirmed as molecular ion peak of ethyl 4-hydroxypentanoate, it may be the ion fragment (ethyl valerate) derived from ethyl 4-hydroxypentanoate. GC-TCD shows that an obvious H_2 peak was found in dehydrogenation gas in Fig. 3.8b, which verifies that the path way of CTH is the metal hydride route which is different from direct hydrogen transfer (Meerwein-Ponndorf-Verley, MPV mechanism) (Gilkey & Xu, 2016). H_2 pressure was calculated by the obtained concentration of acetone and the state equation of ideal gas ($PV = nRT$, $T = 298$ K).

In order to highlight the catalytic activity of A-SCo, Co powder, Ni powder, Al powder and CoAl alloy were also tested (entries 3–6). However, no EL, i-PrOH conversion and acetone production took place during the reaction. In addition, due

Table 3.1 Catalyst screening for GVL production from EL[a]

Entry	Catalyst	EL conversion (%)	*i*-PrOH conversion (%)	GVL yield (%)	H_2 yield (calculated pressure, MPa)
1	Blank	0	0	0	0
2	A-SCo	99.2	4.2	68.0	0.68
3	Co powder	0	0	0	0
4	Ni powder	0	0	0	0
5	Al powder	0	0	0	0
6	CoAl alloy	0	0	0	0
7	Co_2O_3	0	0	0	0
8	CoO	0	0	0	0
9	NiO	0	0	0	0
10	Al_2O_3	0	0	0	0
11	HZSM-5	8.7	0.2	0	0.03
12	A-SCo + HZSM-5	99.2	5.5	94.6	0.89
13	A-SCo + ZnO/HZSM-5	99.5	4.6	91.6	0.72

Reprinted from Elsevier (Ren, 2022)
[a] Reaction conditions: 0.2 mmol EL, 20 mg catalyst, 20 mg solid acid, 2 mL *i*-PrOH, 140 °C, 3 h

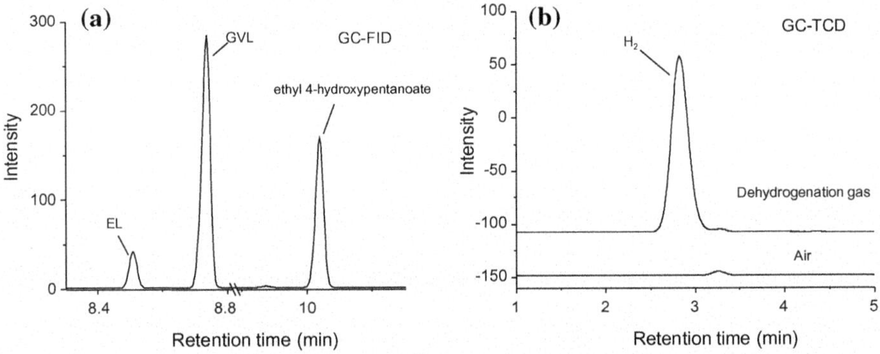

Fig. 3.8 **a** GC-FID of liquid samples catalyzed by A-SCo, reaction condition: 0.2 mmol EL, 20 mg catalyst, 2 mL i-PrOH, 140 °C, 3 h; **b** GC-TCD of dehydrogenation gas catalyzed by A-SCo. Reprinted from Elsevier (Ren, 2022)

to the A-SCo was pretreated with BA process, Co, Ni and Al oxides formed on the surface of A-SCo. Hence, Co_2O_3, CoO, NiO, and Al_2O_3 were also investigated. As is expected, no catalytic activity of these catalyst is exhibited (entries 7–10). It might be demonstrated that the skeletal structure and the distribution of CoAl might impact the EL conversion and i-PrOH dehydrogenation.

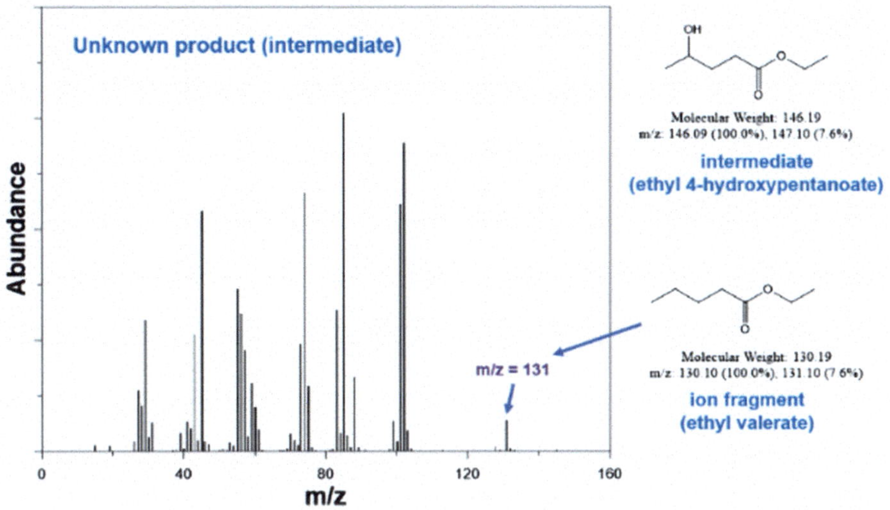

Fig. 3.9 The mass spectrometry data of unknown product (intermediate). Reprinted from Elsevier (Ren, 2022)

According to previous studies (Gilkey & Xu, 2016; Li et al., 2017), Lewis and Brønsted acid sites on the catalyst play important roles for accelerating LA and its ester to form ethyl 4-hydroxypentanoate and produce GVL. Therefore, two types of solid acid HZSM-5 and ZnO/HZSM-5 were investigated. Existing HZSM-5 alone in the reaction cannot activate EL to produce GVL (entry 11). When A-SCo was added with HZSM-5, the EL was almost consumed and the yield of GVL reaches up to 94.6% (entry 12). It seems that HZSM-5 can accelerate the conversion of intermediate into GVL. ZnO/HZSM-5 with higher Lewis/Brønsted acid ratio than HZSM-5 (Ren et al., 2009) shows similar effect to facilitate the GVL production, and 91.6% yield of GVL was obtained (entry 13). High amount of Brønsted acid is more benefit than Lewis acid for ethyl 4-hydroxypentanoate transformation, which is in agreement with previous report (Li et al., 2017).

3.3.3 *The Effect of Various Parameters*

The effect of catalyst loading was first investigated in Table 3.2. EL was almost consumed to afford the 90% yield of GVL when loading of A-SCo ranging from 10 to 40 mg (entries 1–4). It is indicated that the A-SCo shows the good performance at low catalyst loading. The results of i-PrOH conversion and GVL yield show that high loading of solid acid (HZSM-5) not only facilitates the transformation from intermediate into GVL, but also enhances the dehydrogenation of i-PrOH (entries 1, 5–7).

Table 3.2 The effect of A-SCo and HZSM-5 loading

Entry	A-SCo (mg)	HZSM-5 (mg)	EL conversion (%)	*i*-PrOH conversion (%)	GVL yield (%)	H_2 yield (calculated pressure, MPa)
1	10	20	99.4	5.4	94.4	0.88
2	20	20	99.2	5.5	94.6	0.89
3	30	20	> 99.9	5.3	95.0	0.86
4	40	20	> 99.9	5.5	96.2	0.90
5	10	10	98.9	4.8	78.3	0.77
6	10	30	99.6	6.9	96.2	1.11
7	10	40	99.7	7.9	93.7	1.28

Reprinted from Elsevier (Ren, 2022)
Reaction conditions: 0.2 mmol EL, 2 mL *i*-PrOH, 140 °C, 3 h.

Reaction temperature and time effect of EL conversion was also investigated. As shown in Figs. 3.10a and 3.11a, EL converted rapidly at 140 °C and 94% of EL conversion was obtained at the first 0.5 h, but only 66.1% yield of GVL was obtained. It is found that an obvious peak appeared at retention time ranging from 10.0 to 10.1 min on GC chromatogram which is considered as ethyl 4-hydroxypentanoate. After 1 h, EL was almost consumed and the yield of GVL reached to 80.7%, and at the same time the peak of ethyl 4-hydroxypentanoate diminished gradually which suggested the intermediate identity. The highest yield of GVL was 95.6% after 4 h reaction, only trace amount of EL and ethyl 4-hydroxypentanoate could be detected by GC-FID.

Lower temperature such as 120 °C was also concerned. In Figs. 3.4b and 3.11b, the reaction rate of EL is much slower than that of 140 °C, and only 68.1% EL was converted and 39.7% yield of GVL was produced at first 30 min. Although the conversion of EL surpassed 90% after 1 h, the peak of ethyl 4-hydroxypentanoate diminished slowly. Therefore, the final yield of GVL at 120 °C is much lower than

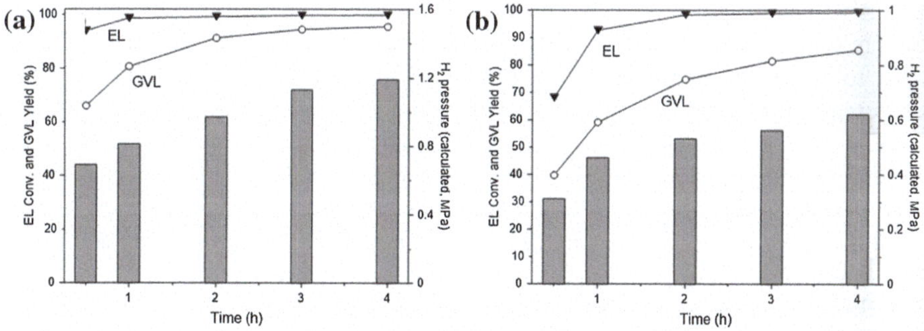

Fig. 3.10 The effect of reaction time for EL conversion and GVL production at **a** 140 °C and **b** 120 °C (0.2 mmol EL, 10 mg A-SCo, 30 mg HZSM-5, 2 mL *i*-PrOH). Reprinted from Elsevier (Ren, 2022)

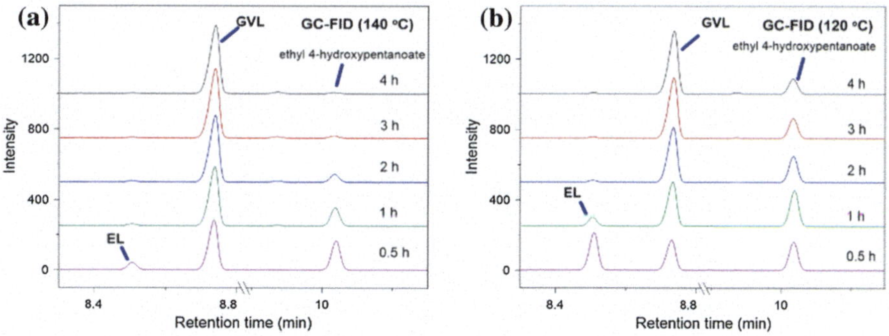

Fig. 3.11 The Gas chromatograms of EL conversion and GVL production with different time at **a** 140 °C and **b** 120 °C. Reprinted from Elsevier (Ren, 2022)

140 °C. It is indicated that higher temperature could accelerate the conversion of intermediate to achieve higher yield of GVL.

3.4 The Availability of Air-Pretreated Process and the Stability of SCo Catalytic Activity

Because pyrophoric character of skeletal metal might be attributed to high active adsorbed H_2, and highly oxidizable of metallic Co and Al (Fouilloux, 1983). As maintained above, BA process and vacuum treatment inhibit the self-ignite process of SCo which can completely oxidize the metallic Co, Ni and Al. According to XPS results, a certain amount of metallic Co, Ni and Al existed on the surface of V-SCo and A-SCo, which indicates that the two treatments might eliminate the pyrophoric H_2 and other reducing agents to obtain a nonpyrophoric A-SCo. A hypothesis is proposed that BA pretreatment creates a low oxygen condition, which consumes the active H_2 on the surface of catalyst. The vacuum treatment pumps out a large amount of easy-desorption H_2, which restrains spontaneous combustion of SCo. In order to verify the hypothesis, Fresh-SCo was purged with N_2 and CO_2 at room temperature for 12 h, respectively, to partly diminish the active H_2 poorly adsorbed on SCo surface, which are named as N-SCo and C-SCo. As expected, all of them lost the pyrophoric character. This indicates that adsorbed H_2 on SCo surface plays an important role for pyrophoricity, BA process can consume the adsorbed H_2 and avoid the spontaneous combustion. It is worth noting that air pretreatment is faster (less than 5 min) and easier handle than that of vacuum process.

Dehydrogenation of i-PrOH was examined in the presence of A-SCo, V-SCo, and C-SCo. Acetone as dehydrogenation product was detected after 30 min, and the use of three of SCo give the similar peak areas of acetone in Fig. 3.12. This indicates that the partial oxidation of SCo surface by the BA process does not decrease the dehydrogenation ability. Evidential experiments were also conducted to investigate the catalytic activities of A-SCo, Fresh-SCo, S-SCo, V-SCo, N-SCo and C-SCo for EL conversion as shown in Tables 3.3 and 3.4.

A-SCo showed great performance for EL conversion, and 96.2% yield of GVL was obtained (entry 1). In comparison, Fresh-SCo, a skeletal Co which is dried and sealed into reactor in glove box under N_2 atmosphere, provided 95.9% yield of GVL and 99.5% conversion of EL (entry 2). It seems that the catalytic activity of skeletal Co is almost unchanged via BA pretreatment. And V-SCo, N-SCo and C-SCo also exhibit comparable yield of GVL (Table 3.4). However, S-SCo does not promote the EL conversion (entry 3). These results agree with characters of Fresh-SCo, A-SCo and S-SCo that A-SCo maintains high surface area (over 27 m^2/g) and plenty of metallic Co on the surface of catalysts which offers more active sites than S-SCo for CTH. The stability of A-SCo in air was also considered. After exposed in air at room temperature for one week, A-SCo still give effective performance for GVL production to obtain 96.8% EL conversion and 92.9% GVL yield (entry 4). The

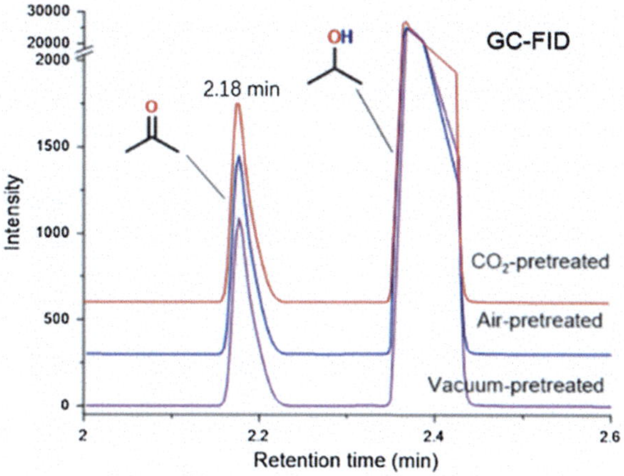

Fig. 3.12 GC-FID of liquid samples catalyzed by A-SCo, V-SCo, and C-SCo, reaction condition: 2 mL *i*-PrOH, 30 mg catalyst, 140 °C, 30 min. Reprinted from Elsevier (Ren, 2022)

Table 3.3 The influence of skeletal Co pretreatment for conversion of EL to GVL[a]

Entry	Pretreatment	EL conversion (%)	GVL yield (%)	H_2 yield (calculated pressure, MPa)
1	A-SCo	99.6	96.2	1.11
2	Fresh-SCo	99.5	95.9	1.08
3	S-SCo	0	0	0.03
4[b]	A-SCo	96.8	92.9	1.27

Reprinted from Elsevier (Ren, 2022)
[a] Reaction conditions: 0.2 mmol EL, 10 mg catalyst, 30 mg HZSM-5, 2 mL *i*-PrOH, 140 °C, 3 h.
[b] A-SCo was exposed in air at room temperature for one week

Table 3.4 The influence of skeletal Co pretreatment (vacuum, N_2 and CO_2) for conversion of EL to GVL

Entry	Pretreatment	EL conversion (%)	GVL yield (%)
1	V-SCo	99.3	95.3
2	N-SCo	99.6	96.1
3	C-SCo	99.2	93.8

Reprinted from Elsevier (Ren, 2022)
Reaction conditions: 0.2 mmol EL, 10 mg catalyst, 30 mg HZSM-5, 2 mL *i*-PrOH, 140 °C, 3 h.

catalyst reuse tests show in Fig. 3.13 that A-SCo can be recycled at least three times and keep the yield of GVL around 50%.

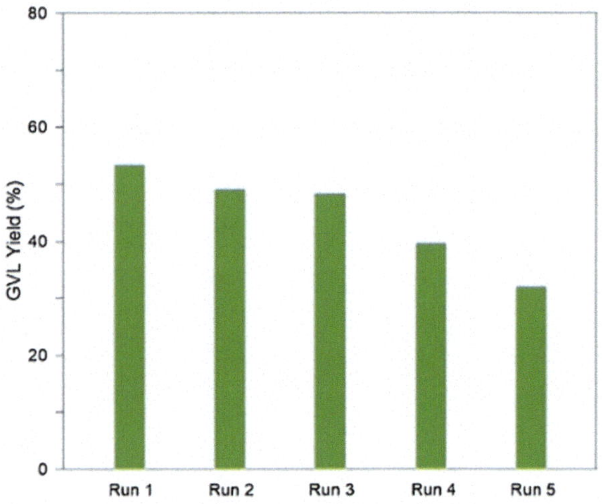

Fig. 3.13 The catalyst reuse stability (0.8 mmol EL, 10 mg A-SCo, 30 mg HZSM-5, 2 mL i-PrOH, 140 °C, 3 h). Reprinted from Elsevier (Ren, 2022)

3.5 The Mechanism of EL CTH Process Over A-SCo

As mentioned above, both the A-SCo and V-SCo containing metallic Co and Al on the surface were detected by XPS. In addition, it is also found that the completed oxidation SCo could not catalyze the dehydrogenation of i-PrOH (entry 3, Table 3.3). This indicates that metallic Co is the catalytic active site for i-PrOH dehydrogenation. In general, Meerwein—Ponndorf—Verley (MPV) and metal hydride routes are two main mechanisms for CTH process when alcohol as hydrogen donors. Because i-PrOH dehydrogenation products of acetone and H_2 were detected during the reaction, the proposed mechanism of EL CTH process over A-SCo are suggested to follow the metal hydride route as shown in Fig. 3.14 according to GC results and previous studies (Gilkey & Xu, 2016; Li et al., 2017). Therefore, i-PrOH is adsorbed first on the catalyst surface. H atoms in hydroxyl group and α-C of i-PrOH are transferred to the surface of A-SCo. Thus, metal-H active species are formed for EL hydrogenation and coupling to release H_2. The only intermediate observed during the reaction is regarded as ethyl 4-hydroxypentanoate due to the GC–MS result, which suggests the acetyl group on EL molecule is hydrogenated by metal-H active species to give intermediate. Then, intramolecular nucleophilic attack takes place to obtain GVL which facilitates by Brønsted acid sites rich H-ZSM-5.

3.6 Conclusions

In summary, the BA process not only avoided spontaneous combustion of air sensitive SCo, but also maintained the catalytic activity of SCo for EL CTH process. The surface structure of SCo maintained after the fast air pretreatment. The highest 96%

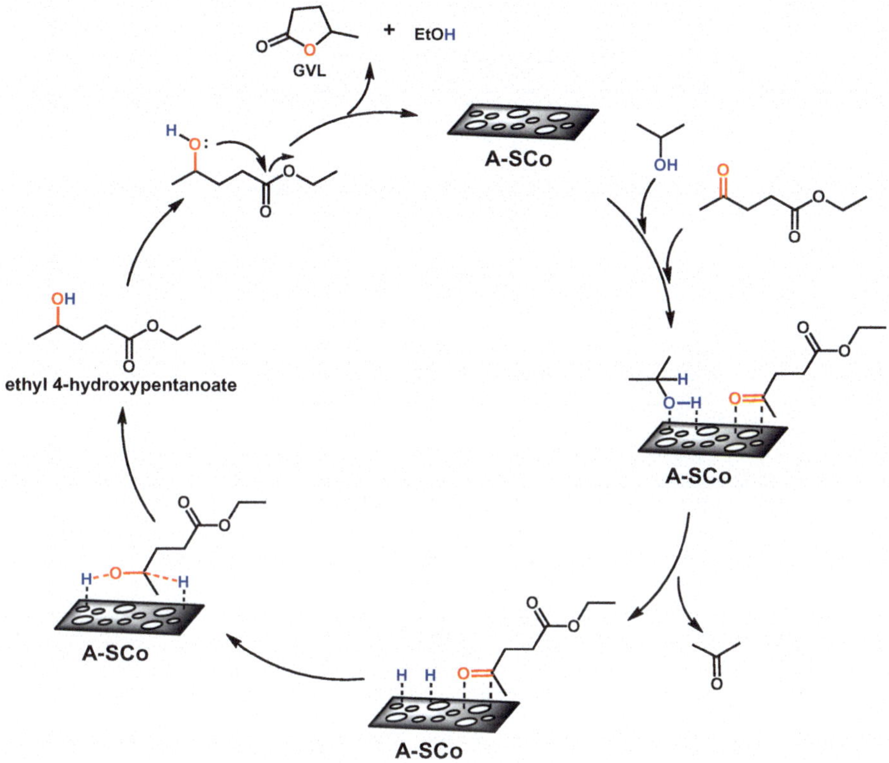

Fig. 3.14 The proposed pathways for the hydrogenation of EL to GVL over air stable skeletal Co catalyst. Reprinted from Elsevier (Ren, 2022)

yield of GVL was obtained in the presence of A-SCo and HZSM-5 at 140 °C. Eliminated the adsorbed H_2 on SCo surface by BA pretreatment contributed to the formation of non-pyrophoric SCo. Transition metals in SCo afford plenty of active site for dehydrogenation of i-PrOH and then hydrogenation of EL to GVL. This work provides a fast and low-cost passivation method to prepare a non-pyrophoric active skeletal metal catalyst for CTH and hydrogenation process.

References

Amarasekara, A. S., & Hasan, M. A. (2015). Pd/C catalyzed conversion of levulinic acid to γ-valerolactone using alcohol as a hydrogen donor under microwave conditions. *Catalysis Communications, 60*, 5–7.

Cao, Y., Chen, S. S., Tsang, D. C. W., Clark, J. H., Budarin, V. L., Hu, C., Wu, K. C. W., & Zhang, S. (2020). Microwave-assisted depolymerization of various types of waste lignins over two-dimensional CuO/BCN catalysts. *Green Chemistry, 22*, 725–736.

Chen, M., Wang, X., Yu, Y., Pei, Z., Bai, X., Sun, C., Huang, R., & Wen, L. (2000). X-ray photoelectron spectroscopy and auger electron spectroscopy studies of Al-doped ZnO films. *Applied Surface Science, 158*, 134–140.

Chen, F., Sahoo, B., Kreyenschulte, C., Lund, H., Zeng, M., He, L., Junge, K., & Beller, M. (2017). Selective cobalt nanoparticles for catalytic transfer hydrogenation of N-heteroarenes. *Chemical Science, 8*, 6239–6246.

Feliu, S., & Barranco, V. (2003). XPS study of the surface chemistry of conventional hot-dip galvanised pure Zn, galvanneal and Zn–Al alloy coatings on steel. *Acta Materialia, 51*, 5413–5424.

Fierro, G., Lo Jacono, M., Inversi, M., Dragone, R., & Porta, P. (2000). TPR and XPS study of cobalt–copper mixed oxide catalysts: evidence of a strong Co–Cu interaction. *Topics in Catalysis, 10*, 39–48.

Fouilloux, P. (1983). The nature of raney nickel, its adsorbed hydrogen and its catalytic activity for hydrogenation reactions (review). *Applied Catalysis, 8*, 1–42.

Gilkey, M. J., & Xu, B. (2016). Heterogeneous catalytic transfer hydrogenation as an effective pathway in biomass upgrading. *ACS Catalysis, 6*, 1420–1436.

Gou, X., Okejiri, F., Zhang, Z., Liu, M., Liu, J., Chen, H., Chen, K., Lu, X., Ouyang, P., & Fu, J. (2020). Tannin-derived bimetallic CuCo/C catalysts for an efficient in-situ hydrogenation of lauric acid in methanol-water media. *Fuel Process Technology, 205*.

Guo, H., Gao, R., Sun, M., Guo, H., Wang, B., & Chen, L. (2019). Cobalt entrapped in N, S-codoped porous carbon: Catalysts for transfer hydrogenation with formic acid. *Chemsuschem, 12*, 487–494.

Hammache, S., Goodwin, J. G., & Oukaci, R. (2002). Passivation of a co–Ru/γ-Al_2O_3 Fischer-Tropsch catalyst. *Catalysis Today, 71*, 361–367.

Hengne, A. M., Kadu, B. S., Biradar, N. S., Chikate, R. C., & Rode, C. V. (2016). Transfer hydrogenation of biomass-derived levulinic acid to γ-valerolactone over supported Ni catalysts. *RSC Advances, 6*, 59753–59761.

Horváth, I. T., Mehdi, H., Fábos, V., Boda, L., & Mika, L. T. (2008). γ-Valerolactone—A sustainable liquid for energy and carbon-based chemicals. *Green Chemistry, 10*, 238–242.

Huang, Y.-B., Yang, T., Luo, Y.-J., Liu, A.-F., Zhou, Y.-H., Pan, H., & Wang, F. (2018). A simple and efficient conversion of cellulose to γ-valerolactone through integrated alcoholysis/transfer hydrogenation system using Ru and aluminum sulfate catalysts. *Catalysis Science & Technology, 8*, 6252–6262.

Huber, F., Yu, Z., Lögdberg, S., Rønning, M., Chen, D., Venvik, H., & Holmen, A. (2006). Remarks on the passivation of reduced Cu-, Ni-, Fe-, Co-based catalysts. *Catalysis Letters.*, 110, 211–220.

Kumar, B. S., Amali, A. J., & Pitchumani, K. (2018). Heterogenization of cobalt nanoparticles on hollow carbon capsules: Lab-in-capsule for catalytic transfer hydrogenation of carbonyl compounds. *Molecular Catalysis, 448*, 153–161.

Li, F., France, L. J., Cai, Z., Li, Y., Liu, S., Lou, H., Long, J., & Li, X. (2017). Catalytic transfer hydrogenation of butyl levulinate to γ-valerolactone over zirconium phosphates with adjustable Lewis and Brønsted acid sites. *Applied Catalysis b: Environmental, 214*, 67–77.

Ni, J., Leng, W., Mao, J., Wang, J., Lin, J., Jiang, D., & Li, X. (2019). Tuning electron density of metal nickel by support defects in Ni/ZrO_2 for selective hydrogenation of fatty acids to alkanes and alcohols. *Applied Catalysis b: Environmental, 253*, 170–178.

Pileidis, F. D., & Titirici, M. M. (2016). Levulinic acid biorefineries: New challenges for efficient utilization of biomass. *Chemsuschem, 9*, 562–582.

Qi, L. & Horváth, I. T. (2012). Catalytic Conversion of Fructose to γ-Valerolactone in γ-Valerolactone. *ACS Catalysis, 2*, 2247–-2249.

Ren, Y., Zhang, F., Hua, W., Yue, Y., & Gao, Z. (2009). ZnO supported on high silica HZSM-5 as new catalysts for dehydrogenation of propane to propene in the presence of CO_2. *Catalysis Today, 148*, 316–322.

Ren, D., Zhao, C., Zhang, N., Norinaga, K., Zeng, X., & Huo, Z. (2022). Catalytic transfer hydrogenation of ethyl levulinate into γ-valerolactone over air-stable skeletal cobalt catalyst. *Journal of Environmental Chemical Engineering, 10(2),* 107–188.

Riva, R., Miessner, H., Vitali, R., & del Piero, G. (2000). Metal–support interaction in Co/SiO_2 and Co/TiO_2. *Applied Catalysis a: General, 196*, 111–123.

Sun, M., Xia, J., Wang, H., Liu, X., Xia, Q., & Wang, Y. (2018). An efficient Ni_xZr_yO catalyst for hydrogenation of bio-derived methyl levulinate to γ-valerolactone in water under low hydrogen pressure. *Applied Catalysis b: Environmental, 227*, 488–498.

Suresh Kumar, B., Puthiaraj, P., Amali, A. J., & Pitchumani, K. (2017). Ultrafine bimetallic PdCo alloy nanoparticles on hollow carbon capsules: An efficient heterogeneous catalyst for transfer hydrogenation of carbonyl compounds.*ACS Sustainable Chemistry & Engineering, 6*, 491–500.

Wang, J., Nie, R., Xu, L., Lyu, X., & Lu, X. (2019). Catalytic transfer hydrogenation of oleic acid to octadecanol over magnetic recoverable cobalt catalysts. *Green Chemistry, 21*, 314–320.

Wolf, M., Fischer, N., & Claeys, M. (2016). Effectiveness of catalyst passivation techniques studied in situ with a magnetometer. *Catalysis Today, 275*, 135–140.

Xu, C., Ouyang, W., Munoz-Batista, M. J., Fernandez-Garcia, M., & Luque, R. (2018). Highly active catalytic ruthenium/TiO_2 nanomaterials for continuous production of gamma-valerolactone. *Chemsuschem, 11*, 2604–2611.

Yang, Z., Huang, Y. B., Guo, Q. X., & Fu, Y. (2013). RANEY(R) Ni catalyzed transfer hydrogenation of levulinate esters to gamma-valerolactone at room temperature. *Chemical Communications, 49*, 5328–5330.

Yang, Y., Xu, X., Zou, W., Yue, H., Tian, G., & Feng, S. (2016). Transfer hydrogenation of methyl levulinate into gamma-valerolactone, 1,4-pentanediol, and 1-pentanol over Cu–ZrO_2 catalyst under solvothermal conditions. *Catalysis Communications, 76*, 50–53.

Yuan, M., Long, Y., Yang, J., Hu, X., Xu, D., Zhu, Y., & Dong, Z. (2018). Biomass sucrose-derived cobalt@Nitrogen-doped carbon for catalytic transfer hydrogenation of nitroarenes with formic acid. *Chemsuschem, 11*, 4156–4165.

Zhang, G., Sun, S., Bostetter, M., Poulin, S., & Sacher, E. (2010). Chemical and morphological characterizations of CoNi alloy nanoparticles formed by co-evaporation onto highly oriented pyrolytic graphite. *Journal of Colloid and Interface Science, 350*, 16–21.

Zhang, Z., Yang, Q., Chen, H., Chen, K., Lu, X., Ouyang, P., Fu, J., & Chen, J. G. (2018). In situ hydrogenation and decarboxylation of oleic acid into heptadecane over a Cu–Ni alloy catalyst using methanol as a hydrogen carrier. *Green Chemistry, 20*, 197–205.

Zhou, H., Song, J., Fan, H., Zhang, B., Yang, Y., Hu, J., Zhu, Q., & Han, B. (2014). Cobalt catalysts: Very efficient for hydrogenation of biomass-derived ethyl levulinate to gamma-valerolactone under mild conditions. *Green Chemistry, 16*, 3870–3875.

Chapter 4
Boosting In-Situ Hydrodeoxygenation of Fatty Acids Over a Fine and Oxygen-Vacancy-Rich NiAl Catalyst

4.1 Introduction

The attitude of world-wide reduction in carbon emissions has reached a broad agreement in academic and industry communities. Therefore, adopting carbon–neutral and renewable biomass as substitutes for fossil resources become a research hot spot in the past few decades (Lu et al., 2021). Bio-derived lipids including microalgae oil, waste cooking oil with huge production and high organic carbon contents are promising alternative feedstocks for biofuels manufacture (Chen et al., 2011). To boost the energy density and market value of lipids, removing oxygen and reserving long-chain alkanes structure of fatty acid via catalytic hydrodeoxygenation (HDO) process to prepare hydrocarbon aviation fuels at high temperature has been widely studied (Yao et al., 2021). Precious metal catalysts, such as Pt (Kon et al., 2014), Pd (Sun et al., 2015), Ru (Ali & Zhao, 2020), afford superior activity for HDO of fatty acids at relatively mild condition (<300 °C) (Ding et al., 2022). However, given the high cost of precious metals, scientists are being motivated to exploit non-precious metals, especially Ni-based materials, to cut costs (Chen et al., 2015; Jiraroj et al., 2021).

Monometallic Ni tends to result in the cleave of C–C bond in alkyl chain at high temperature and is easy to deposit carbon, which reducing the product selectivity and life of catalyst (Zhang et al., 2018b). In this light, adding a second (or third) metal to afford Ni-based alloys or constructing appropriate supports are effective avenues for regulating the activity and stability of catalysts. For instances, bimetallic CuNi alloy catalyst was prepared and achieved 92.7% yield of n-heptadecane at 330 °C with methanol as hydrogen donor (Li et al., 2015). The formation of CuNi alloy exhibited strong adsorption of carboxylic acid and inhibited the cracking of C–C bond (Zhang et al., 2018a). 91.3% n-heptadecane yield was reached over a well prepared trimetallic CuNi/CoOx at 240 °C (Wang et al., 2020a). The authors suggested a synergetic effect between Cu, Ni and CoOx which facilitates the HDO process of oleic acid. Palygorskite was reported as a cheap and effective substrate to give supported Ni catalyst which afforded over 80% yield of n-alkanes at 310 °C

Z. Huo, *Diverse Hydrogen Sources for Biomass-derivatives Conversion*,
https://doi.org/10.1007/978-981-99-1673-3_4

with 4 MPa H_2 (Lycourghiotis et al., 2019). Although some of them achieved good yield of desired alkanes, conducting fatty acids via HDO process liquid hydrogen donor at mild condition over Ni-based catalysts remains a challenging subject.

As is well-known, exposing plenty active sites and tailoring the binding strength of reactants on the surface of catalyst are general principles for preparing perfect catalysts. Construction of high loading Ni-based supported catalyst with small particle size would expose large amount of metal/support interfaces which are typically recognized as active sites, but suppressing metallic Ni nanoparticles agglomeration during the preparation process, which would result in catalyst deactivation, is challenging. Additionally, in recent years, defect engineering as versatile strategy has displayed great positive performance to tune the catalytic activity of heterogenous catalyst in hydrogenation reaction (Cui et al., 2020; Jia et al., 2017). Lv et al. developed a defect-rich (oxygen vacancy, OV) NiW bimetallic which afforded 100% yield of total alkanes at 320 °C. Density functional theory (DFT) calculation results demonstrated that the high activity was attributed to the OV sites introduced via partially reduced WO_3 by H_2 which had strong ability to absorb carboxyl group in fatty acid (Xing et al., 2020). Li et al. reported that negatively charged Ni, which was created by electron transfer from t-ZrO_2 support with appropriate OV, rendered 86.03% selectivity of n-heptadecane at 240 °C (Ni et al., 2019). Additionally, the electron transport between Ni and second metal in NiIn/SiO_2 were also observed, which facilitated the HDO process to yield fatty alcohol at 270 °C (Wang et al., 2020b). Based on the previous reports, providing plentiful active sites by high loading and fine Ni catalyst with OV-rich support and tuning the catalyst charge density which highly desired for activating carboxyl group would facilitate the yield of alkanes for HDO of fatty acids.

Herein, fine and high loading Ni-based particles modified with amorphous AlO_x were prepared for catalytic transfer hydrogenation (CTH), one commonly used in-situ HDO technique, of fatty acid into alkane with isopropanol as hydrogen donor in this study. The particle size of catalyst was reduced dramatically though introducing small amount of Al species, which favorably exposes abundant active sites. Highly defective Ni–AlO_x and electron transfer from Ni to AlO_x were appeared to improve HDO reaction. DFT calculation results show that the formation of defective interfaces possessed higher charge density in interface of Ni–AlO_x and stronger adsorption of carboxyl group, to facilitate the HDO of fatty acids. In the end, 93.2% yield of n-heptadecane was obtained via CTH of stearic acid at 250 °C. The catalyst was reused at least 4 times without deactivation.

4.2 Materials and Methods

4.2.1 Experimental Materials

Stearic acid (98%), n-heptadecane (AR), n-octadecane (99%) and Na_2CO_3 (AR) were purchased from Shanghai Macklin Biochemical Co. Ltd., China. Octadecanol (99%), $Al(NO_3)_3·9H_2O$ (AR), NaOH (AR), γ-Al_2O_3 (99%) were purchased from General-Reagent, Shanghai Titan Chemical Co. Ltd., China. Isopropanol (AR), n-hexane (AR) were obtained from Energy Chemical Co. Ltd., China. Acetone was purchased from Sinopharm Chemical Reagent Co. Ltd., China. $Ni(NO_3)_2·6H_2O$ was purchased from Shanghai Lingfeng Chemical Reagent Co. Ltd., China. All the above chemicals were used as received without further purification.

4.2.2 Catalyst Preparation

The Ni–AlO_x catalyst was prepared according to the previous report (He et al., 2012). The typical process was conducted using hydrothermal precipitation and reduction method: an aqueous solution (50 mL) containing $Ni(NO_3)_2·6H_2O$ of 8.18 g and $Al(NO_3)_3·9H_2O$ of 3.52 g (Ni/Al molar ratio equal to 3:1) was labeled as solution I. An aqueous solution (50 mL) containing 4.24 g of Na_2CO_3 (0.04 mol) and 2 g of NaOH (0.05 mol) was labeled as solution II. Then solution II was added dropwise into solution I continually under vigorous stirring with a rate of 100 mL/h. NaOH aqueous solution (3 mol/L) was used to adjust the pH of the suspension to 10. Then, the suspension liquid was added to Stainless steel hydrothermal reactor and kept 65 °C for 18 h to get hydrotalcite-like structure. The precipitate was collected by centrifugation and washed repeatedly with deionized water until the pH value decreased to 7. After drying at 80 °C for 16 h, the precursor was reduced at 500 °C for 5 h in 5% H_2/Ar flow. Other Ni_a–Al_bO_x (n (Ni)/n(Al) = a/b) catalysts were made in the same way, which are named respectively as Ni–Al_2O_x, Ni–Al_1O_x, Ni–$Al_{0.5}O_x$, Ni–$Al_{0.25}O_x$, Ni–$Al_{0.2}O_x$, Ni–$Al_{0.14}O_x$ and Ni powder (There is no Al added in the preparation process).

4.2.3 Typical Experiment and Product Analysis

CTH of stearic acid reaction was carried out in a microbatch reactor (with the volume of 4.67 mL) that was assembled from a 3/8-inch tube and two 3/8-inch caps purchased from Swagelok, USA. In a typical run, 2.5 mL isopropanol, 25 mg stearic acid and desired amount of catalyst were added into the reactor. Then the reactor was sealed and heated at suitable temperature and time. After reaction, the reactor was submerged in cold water to quench the reaction. The liquid products were filtered by a 0.45 μm organic filter head and then analyzed by gas chromatography (GC, Agilent 7890A) with a 30 m capillary column (HP-5) using a flame ionization detector (FID). The catalyst was recovered by magnet and successively washed by isopropanol for 5 times, then used for the next batch of reaction. The GC detecting conditions were as follows: nitrogen was the carrier gas, the injection port temperature was 280 °C, and the detector (FID) temperature was 310 °C. The column was heated from 50 to 250 °C by 10 °C min^{-1}. A quantitative analysis was performed using calibrationurves for every compound in the mixture.

The conversion of stearic acid and yield of products were calculated using the following equations

$$Conversion(\%) = \frac{n_{SA} - n'_{SA}}{n_{SA}} \times 100$$

$$Yield(\%) = \frac{n_P}{n_{SA}} \times 100$$

Here n_{SA} is the initial mol of stearic acid and n'_{SA} is the residual mol of stearic acid after reaction, n_p is the final mol of product.

4.2.4 Density Functional Theory Calculations

Density functional theory (DFT) calculations were carried out using the Vienna Ab-initio Simulation Package (VASP). The spin-polarized projector augmented wave (PAW) method and the PBE electron exchange–correlation functional with the generalized gradient approximation (GGA) were applied in our calculations. The kinetic energy cutoff for the wave function expanded in the plane-wave basis was set as 450 eV. To simplify the calculation, propionic acid was modeled as stearic acid (Kong et al., 2018). AlO_x cluster on top of a 5-layer Ni(111) slab with 4 × 4 supercell was set as the Ni–AlO_x catalyst (Xing et al., 2020). For comparison, 4-layer Ni(111) slab with 4 × 4 supercell was set as the Ni powder, Al_2O_3 cluster on top of a 4-layer Ni(111) slab with 4 × 4 supercell was set as the Ni/Al_2O_3-IMP catalyst. The vacuum space along the z direction was set to be 20 Å to eliminate the interaction between neighboring slabs. The upper two-layer atoms for all models in the cell were allowed to relax during the structure optimization and the bottom three-layer atoms were fixed. The Brillouin-zone integration was performed along with a 4 × 4 × 1 Monkhorst–Pack grid for all the slabs. All geometry optimization was performed until all forces were smaller than 0.03 eV/Å.

4.3 Results and Discussion

4.3.1 Synthesis and Physicochemical Characterizations

In order to fabricate Ni-based catalyst with fine particle size and OV sites. AlO_x modified Ni nanoparticles with a mass of OV sites were prepared via a hydrothermal process and then reduced in H_2/Ar stream at 500 °C as shown in Fig. 4.1a. The as-prepared NiAl hybrids, for example Ni–$Al_{0.33}O_x$, were named by the mole ratio of Ni to Al. Scanning electron microscopy (SEM) and transmission electron microscope (TEM) images display that Ni–$Al_{0.33}O_x$ appears as fine particles with small and uniform size around 5–15 nm (Figs. 4.1b and 4.2a, b, e). High resolution TEM (HRTEM) images (Fig. 4.1c) show Ni–$Al_{0.33}O_x$ particles with defined lattice fringes. The lattice fringes on single nanoparticle with d-spacing of 0.203 nm and 0.209 nm correspond to Ni (111) and NiO (200) plane, respectively (Fig. 4.1d). No Al species are observed indicating the formation of amorphous AlO_x. Selected area electron diffraction (SAED) result with bright rings confirms the coexistence of Ni and NiO in Ni–$Al_{0.33}O_x$ (Fig. 4.1e). Energy dispersive X-ray spectroscopy (EDS) line-profile and elemental mapping of micro and nano scale show the overlap of Ni, Al, O, indicating the homogeneously distribution of these elements (Figs. 4.1f–i and 4.2c, d). Considering the size of Ni–$Al_{0.33}O_x$, EDS line-profile over several particles was conducted and the results exhibit Ni, Al, O with nearly consistent patterns of distribution, suggesting that nanoparticles of Ni–$Al_{0.33}O_x$ are constructed as small Ni, NiO crystals and amorphous AlO_x, resulting in the formation of massive metal-oxide interfaces (Fig. 4.1j). In contrast, Ni/Al_2O_3-IMP prepared wetness impregnation with same ratio of Ni and Al presents larger particle size than that of Ni–$Al_{0.33}O_x$ (Figs. 4.3a, b and 4.4a). Additionally, uneven distribution particle size of impregnation derived Ni/Al_2O_3-IMP is observed in high magnification TEM (Fig. 4.4b). Similarly, HRTEM and SAED also display the presence of Ni and NiO, and no Al crystal species is discovered (Fig. 4.4c, d). EDS mapping shows agglomeration of Ni particles and the same dispersion of Al and O, demonstrating the Ni/Al_2O_3-IMP may contains higher amount of metallic Ni than Ni–$Al_{0.33}O_x$ (Fig. 4.4e–h). It is also confirmed by the EDS line-profile in micro and nano scale (Figs. 4.3c, d and 4.4i). To verify the partial results of microanalysis described above, X-ray diffraction (XRD) tests were carried out to probe the crystal structure of samples. As shown in Fig. 4.1k, Ni and NiO are identified in both Ni–$Al_{0.33}O_x$ and Ni/Al_2O_3-IMP. Impregnation derived Ni/Al_2O_3-IMP shows sharper diffraction peaks than Ni–$Al_{0.33}O_x$, indicating the larger grain size. In addition, comparing with the pure Ni powder, introduction of amorphous Al oxide results the formation of NiO. Moreover, hydrothermal derived Ni–$Al_{0.33}O_x$ has higher NiO content than Ni/Al_2O_3-IMP from the XRD results. The formation of NiO is mainly attributed to the air exposure of Ni catalyst giving rise to the oxidation of metallic Ni nanoparticles. Therefore, a fine and high loading Ni catalyst modifying by AlO_x was fabricated via hydrothermal and reduction treatment.

The surface composition and electronic states of Ni–Al$0_{.33}O_x$ and Ni/Al_2O_3-IMP were explored by X-ray photoelectron spectroscopy (XPS). As shown in Fig. 4.5 a, Ni

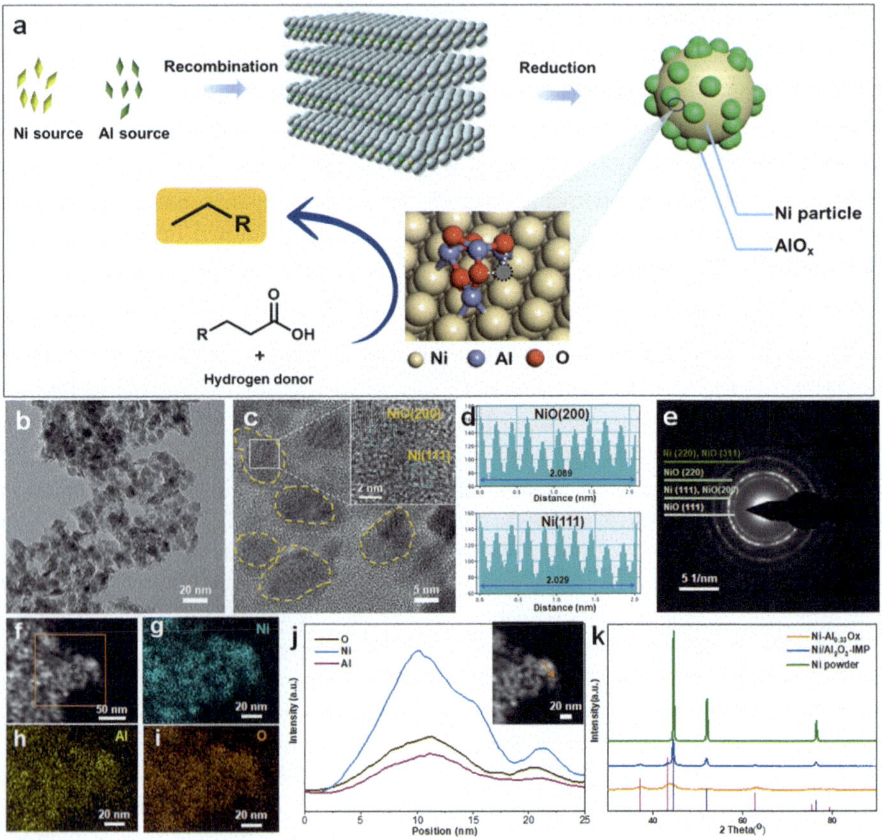

Fig. 4.1 **a** Schematic illustration for the synthesis of Ni–AlOx; **b** TEM image of Ni–$Al_{0.33}O_x$; **c**–**e** HRTEM image, intensity profiles of d-spacing and SAED of Ni–$Al_{0.33}O_x$; **f**–**i** HAADF-STEM image and the corresponding EDS elemental mapping of Ni–$Al_{0.33}O_x$; **j** EDS line-profile over several particles of Ni–$Al_{0.33}O_x$; **k** XRD patterns of Ni–$Al_{0.33}O_x$, Ni/Al_2O_3-IMP and Ni powder

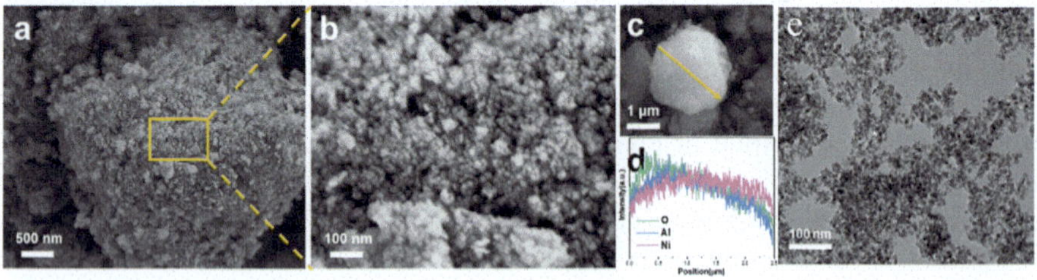

Fig. 4.2 **a** Low and **b** high multiple SEM images of Ni–$Al_{0.33}O_x$; **c**, **d** EDS line-profile over particles of Ni–$Al_{0.33}O_x$; **e** Low multiple TEM image of Ni–$Al_{0.33}O_x$

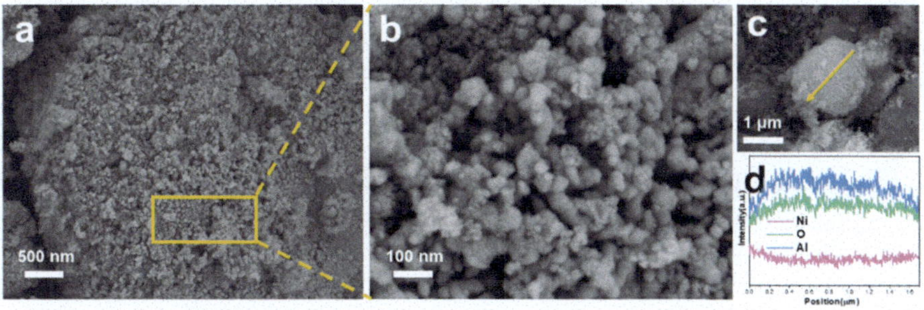

Fig. 4.3 **a** Low and **b** high multiple SEM images of Ni–Al_2O_3-IMP; **c**, **d** EDS line-profile over particles of Ni–Al_2O_3-IMP

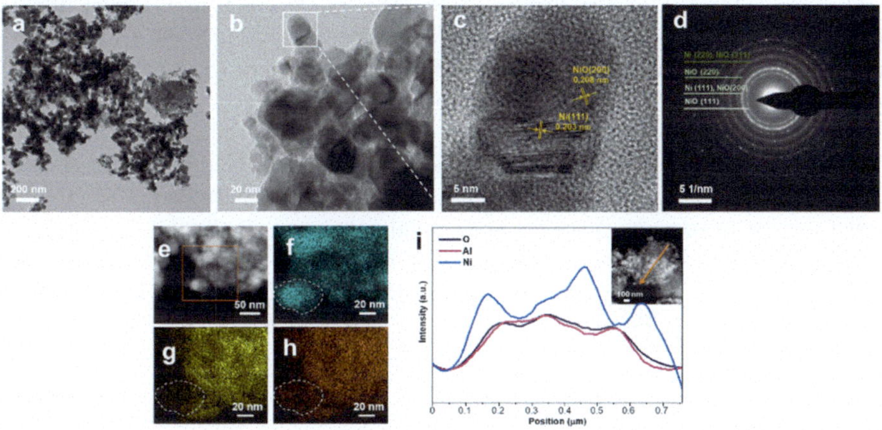

Fig. 4.4 **a**, **b** TEM image of Ni–Al_2O_3-IMP; **c**, **d** HRTEM image, intensity profiles of d-spacing and SAED of Ni–Al_2O_3-IMP; **e–h** HAADF-STEM image and the corresponding EDS elemental mapping of Ni–Al_2O_3-IMP; **i** EDS line-profile of Ni–Al_2O_3-IMP

$2p_{3/2}$ spectrum of Ni–$Al0_{.33}O_x$ can be fitted by four peaks which are assigned to Ni^0 (852.5 eV), Ni^{2+} (854.5 eV), Ni^{3+} (856.1 eV) and shakeup satellite peak (861.3 eV). In effect, the peaks with binding energy values between 851.6 eV and 852.9 eV can be referred as Ni^0 (Chen et al., 2017). It is clearly shown that metallic Ni in Ni–$Al0_{.33}O_x$ possesses higher binding energy than that of Ni/Al_2O_3-IMP, implying the lower electron density of Ni^0 in Ni–$Al0_{.33}O_x$ prepared via hydrothermal Al species modification. Moreover, the peaks assigned as oxidation state of Ni (Ni^{2+}/Ni^{3+}) in Ni–$Al0_{.33}O_x$ also have a positive shift of binding energy with respect to Ni/Al_2O_3-IMP, which illustrates the distinct charge transfer between Ni and Al species on the surface of Ni–$Al0_{.33}O_x$. In addition, the atomic ratio of $Ni^0/(Ni^{2+} + Ni^{3+})$ on the surface of Ni–$Al0_{.33}O_x$ is 7.2% which is obviously lower compared to that in Ni/Al_2O_3-IMP (14.2%), suggesting that the homogeneously distribution of AlO_x promote the electron loss of metallic Ni, which is in accordance with the XRD results. Al 2p and Ni 3p spectra of as-prepared catalysts in Fig. 4.5 b show two main peaks located at 73.8/74.2 eV and 67.6/67.0 eV, respectively. The signals of Al 2p with lower and Ni 3p with higher binding energy for Ni–$Al_{0.33}O_x$ than those for Ni/Al_2O_3-IMP is another evidence for presence of strong interaction between Ni and AlO_x when the NiAl catalyst are synthesized via hydrothermal process. Although Al oxide is reported as an unreducible material, the charge transfer may cause by the Ni ions doping into amorphous AlO_x due to the well dispersed Ni and Al species, which changes the pristine electronic structure (Luo et al., 2021). O 1 s core-level signal of as-prepared NiAl-based catalysts in Fig. 4.5c is a good indicator for oxygen vacancy, which can be fitted to three peaks located around 529.2 and 531.4 eV, three of them assign to lattice oxygen ($O_{lattice}$), oxygen vacancy ($O_{vacancy}$) and adsorbed oxygen (O_{absorb}) species, respectively (Liu et al., 2021; Zhou et al., 2021). As expected, the content of $O_{vacancy}$ (28.8%) represented by the atomic ratio of $O_{vacancy}/(O_{lattice} + O_{vacancy} + O_{absorb})$ on the surface of Ni–$Al_{0.33}O_x$ is much higher than that of Ni/Al_2O_3-IMP (11.1%), which also supports the outcome of Ni 2p, Ni 3p and Al 2p. Apart from the XPS results, electron paramagnetic resonance (EPR) measurements were conducted to investigate oxygen vacancies of as-prepared catalysts. As shown in Fig. 4.5d, three pairs of peaks at a g value of 2.003 are observed, and the intensity of these peaks follows the order Ni–$Al_{0.33}O_x$ > Ni/Al_2O_3-IMP > Ni powder. The sharper and higher peaks imply the higher unsaturated state of catalyst, further indicating the higher density of oxygen vacancies in the catalyst (Cao et al., 2020; Li et al., 2021; Meng et al., 2021). Clearly, introduction of Al oxides on the Ni particle to construct metal/oxides interfaces provides more OV sites, especially for the hydrothermal derived Ni–$Al_{0.33}O_x$. These EPR results give strong evidence for confirming the XPS results.

4.3.1.1 Investigation of Catalytic Performance

In order to verify the active sites for Ni–Al based catalysts, Ni powder, Ni/Al_2O_3-IMP, Ni–$Al_{0.33}O_x$ and Ni–$Al_{0.33}O_x$-air were first investigated for the HDO of stearic acid at 250 °C using isopropanol as hydrogen donor. As shown in Fig. 4.6a, Ni powder displays low catalytic activity, giving only trace amount of n-heptadecane and 11.4% yield of octadecanol. Although over 99% feedstock was converted, most of them reacted with isopropanol to generate relative ester, which is consistent with previous report (Wang et al., 2019). Octadecanol and ester are referred as important intermediates for fatty acids derived alkane production. In general, metallic Ni is regarded as a good material for HDO reaction, but large particle size of Ni powder has few active sites available. Additionally, 78.1% yield of n-heptadecane was obtained when Ni/Al_2O_3-IMP was loaded, however, the remaining 20.2% intermediate ester was not converted. In contrast to Ni powder, Ni/Al_2O_3-IMP possessing smaller particle size and higher content OV sites after metal/oxide interfaces were constructed, suggesting that catalytic activity related to the particle size and OV content of catalyst. Note that no desired alkane but 4.5% yield of octadecanol are observed when $Al_{0.33}O_x$-air was used. XRD patten of Ni–$Al_{0.33}O_x$-air in Fig. 4.5e shows that only NiO was detected,

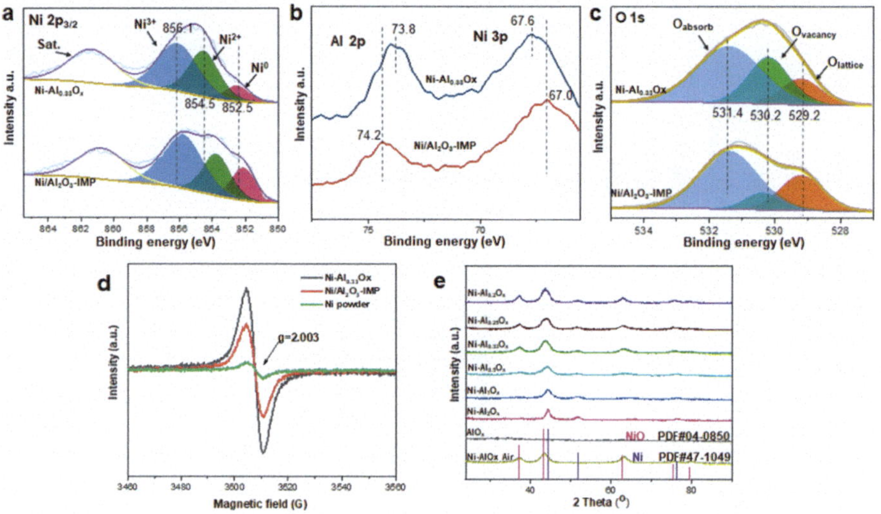

Fig. 4.5 XPS spectra of **a** Ni 2p3/2, **b** Al 2p, Ni 3p, **c** O 1 s; **d** EPR spectra and **e** XRD patterns of catalysts

which demonstrates the catalytic activity of Ni oxide is poor, implying the metallic state of Ni nanoparticles plays a vital role for HDO of fatty acids. Although the ratio of Ni0 on the surface of $Al_{0.33}O_x$ is obviously lower than that of Ni/Al_2O_3-IMP according to the $Ni2p_{3/2}$ and XPS survey spectrum results (Figs. 4.5a, 4.7 and Table 4.1), $Al_{0.33}O_x$ still exhibits 93.2% yield of n-heptadecane and small amount ester production which is the best result among these catalysts. These results further infer that small and OV-rich NiAl catalyst possesses higher activity, and strong charge transfer between Ni^0 and AlO_x in Ni–$Al_{0.33}O_x$ may boosts the stearic acid HDO process, even the amount of Ni^0 is low.

Further, the loading of Al species in Ni–AlO_x was regulated from 0.2 to 2 based on the atomic ratio of Al/Ni to investigate the effect for HDO of stearic acid. As shown in Fig. 4.6a, Ni–$Al_{0.2}O_x$ with small dosage of Al affords 76.9% yield of n-heptadecane. With the dosage of Al increased, the desired alkane yield improved significantly, until ratio of Ni/Al reached 1/0.33. XRD patterns of as-prepared catalysts in Fig. 4.5e show that the wide XRD peaks of Ni–$Al_{0.2}O_x$ and Ni–$Al_{0.25}O_x$ resembling Ni–$Al_{0.33}O_x$ demonstrate the existence of small particle size of Ni and NiO, implying the excessive loading of Ni nanoparticles lacking sufficient support from AlO_x cannot exhibit the maximum catalytic performance. Further rises in the proportion of Al would lead to a rapid decrease in n-heptadecane yield. The signal of NiO diffraction peaks gradually disappears with the raising of Al/Ni ratio, which may attribute to the high loading of AlO_x blocking the interaction between metallic Ni and atmosphere. For comparison, only 80.1% yield of n-heptadecane was obtained over the Ni–$Al_{0.5}O_x$. The morphology features of Ni–$Al_{0.5}O_x$ from TEM, HRTEM and SAED information in Fig. 4.8 display that particle size of Ni particle is comparable to that of Ni–$Al_{0.33}O_x$.

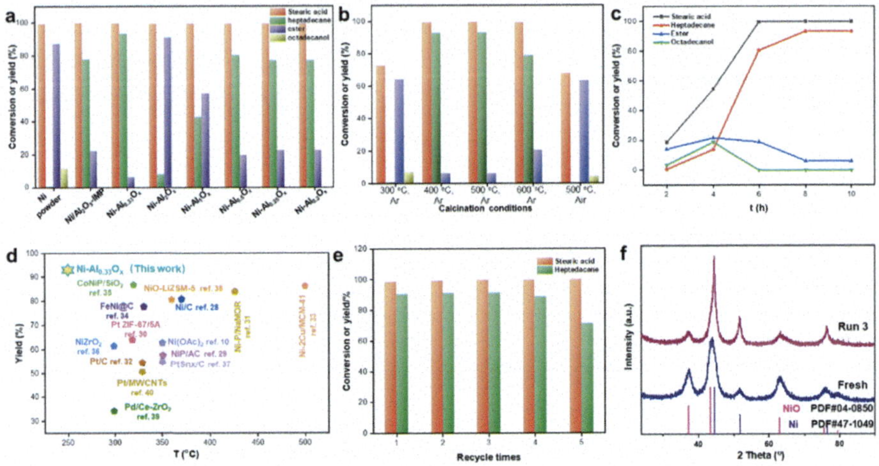

Fig. 4.6 The effect of **a** different Ni-based catalysts and **b** reduction temperature of Ni–$Al_{0.33}O_x$ for CTH of stearic acid (Reaction condition: 25 mg stearic acid, 10 mg catalyst, 2.5 mL isopropanol, 250 °C, 8 h), **c** products distribution of CTH process over Ni–$Al_{0.33}O_x$ at 250 °C with different reaction time, **d** activities of Ni and other (non-) precious catalysts in references and Ni–$Al_{0.33}O_x$ in this work, **e** stability tests of Ni–$Al_{0.33}O_x$ (Reaction condition: 25 mg stearic acid, 10 mg catalyst, 2.5 mL isopropanol, 250 °C, 8 h), **f** XRD patterns of fresh and recycled Ni–$Al_{0.33}O_x$ (used 3 times)

Fig. 4.7 XPS survey spectra of Ni–$Al_{0.33}O_x$, Ni/Al_2O_3-IMP

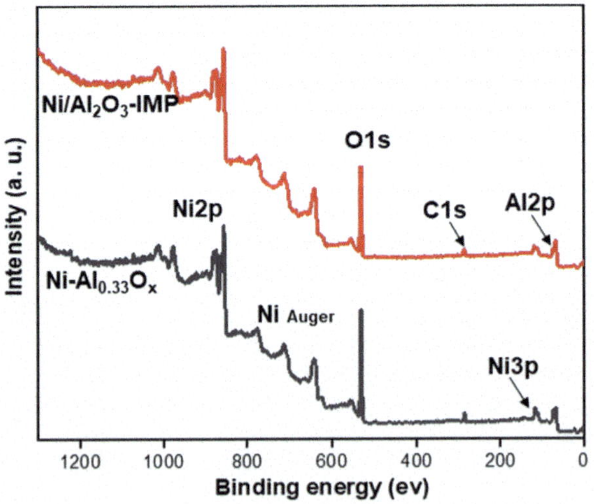

Table 4.1 XPS survey spectra of Ni–$Al_{0.33}O_x$, Ni/Al_2O_3-IMP

	Ni (mol %)	Al (mol %)	O (mol %)
Ni–$Al_{0.33}O_x$	23.1	13.8	63.1
Ni/Al_2O_3-IMP	27.9	7.7	64.4

In addition, although the binding energies of Ni $2p_{3/2}$, Ni 3p and Al 2p for two catalysts shown in Fig. 4.9a, b are very close, indicating that the change of Al/Ni ratio would not affect the charge transfer between Ni and AlO_x, but higher amount of AlO_x suppress the oxidation of Ni particles. Meanwhile, the content of $O_{vacancy}$ in Ni–$Al_{0.5}O_x$ is 19.5%, which is lower than Ni–$Al_{0.33}O_x$ (Fig. 4.9c). Therefore, high dosage of Al favorably maintains the small particle size and metallic state of Ni, but reduces the loading of Ni and density of OV sites, which contribute the inferior catalytic activity.

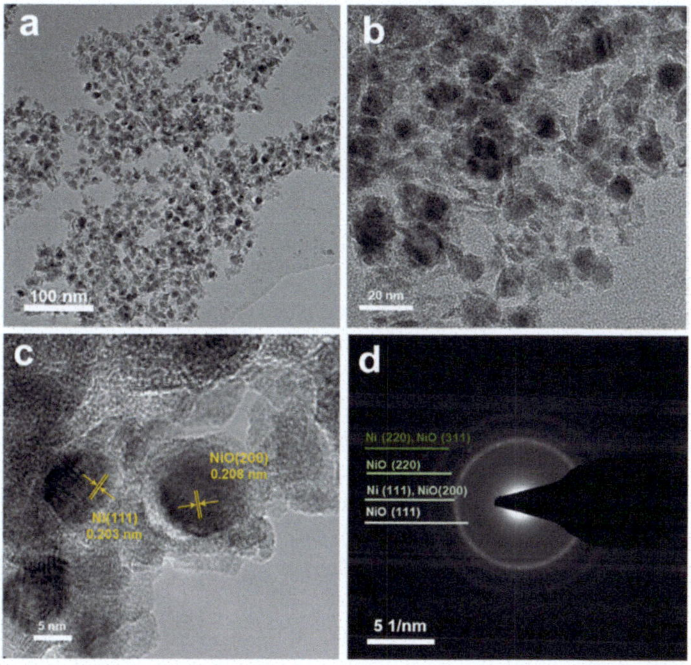

Fig. 4.8 **a** Low and **b** high multiple TEM images and of Ni–$Al_{0.5}O_x$; **c, d** HRTEM image, intensity profiles of d-spacing and SAED of Ni–$Al_{0.5}O_x$

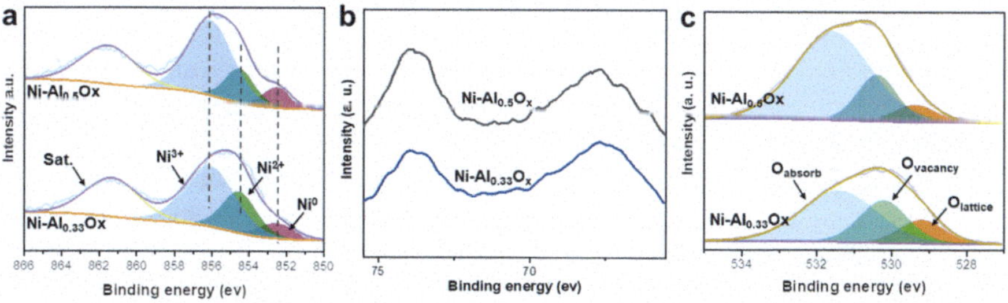

Fig. 4.9 XPS spectra of **a** Ni $2p_{3/2}$, **b** Al 2p, Ni 3p, **c** O 1 s for Ni–$Al_{0.33}O_x$ and Ni–$Al_{0.5}O_x$

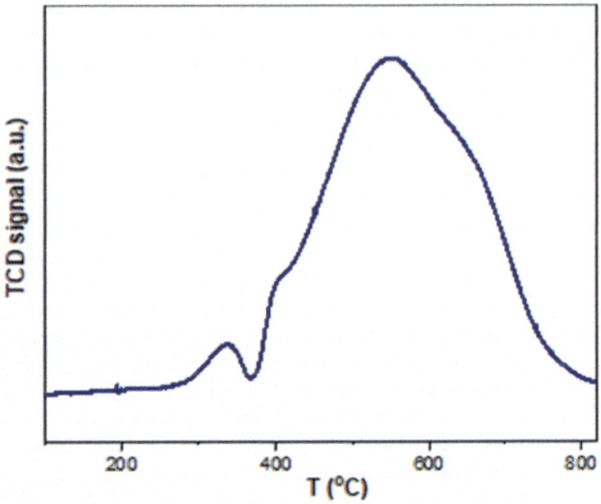

Fig. 4.10 H_2-TPR profile of Ni–$Al_{0.33}O_x$

The influence of catalyst reduction temperature from 300 to 600 °C for Ni–$Al_{0.33}O_x$ performance was explored. As shown in Fig. 4.6b, 400 and 500 °C are preferable temperatures for achieving over 90% yield of n-heptadecane. Lower (300 °C) or higher (600 °C) reduction temperature reduce the catalytic activity and only afford 0.8% and 79.0% n-heptadecane yield, respectively. H_2-TPR measurement displays two broad reduction peaks ranging from 300 to 765 °C (Fig. 4.10), revealing that the as-prepared catalyst treated at 300 °C in H2/Ar stream are hardly reduced, which confirms the failing performance of the oxidation state of Ni. Meanwhile, diffraction peaks of Ni from XRD patterns became sharper in Ni–$Al_{0.33}O_x$ when the reduction temperature reached to 600 °C (Fig. 4.11), indicating the particle sintering, which may contribute the weak activity of catalyst.

The catalyst loading, reaction temperature and time for HDO of stearic acid over Ni–$Al_{0.33}O_x$ were also optimized. 10 mg catalyst is appropriate amount at 250 °C and excessive dosage cannot improve the n-heptadecane yield (Fig. 4.12). The distribution of feedstock and products over reaction time at 250 °C shows in Fig. 4.6c and Fig. 4.13. As expected, n-heptadecane yield increases with time, and the highest yield was achieved at 8 h. In addition, octadecanol and ester behaving as intermediates for the yield first elevated and then declined. Moreover, the production of n-heptadecane was suppressed when 200 and 230 °C were used (Figs. 4.14 and 4.5.15). Compared with other reported precious metals or Ni-based catalysts in recent years, Ni–$Al_{0.33}O_x$ displays excellent catalytic performance for high yield of desired alkane production at mild condition (Fig. 4.6d) (Crawford & Carreon, 2018; Fu et al., 2015; Li et al., 2020; Miao et al., 2016; Rakmae et al., 2020; Shim et al., 2018; Wu et al., 2016; Xin et al., 2019; Yang et al., 2013; Yeh et al., 2015; Zhang et al., 2018b, 2019; Zheng et al., 2020; Zhong et al., 2021).

Stability of as-prepared Ni–$Al_{0.33}O_x$ was evaluated at optimized condition. As shown in Fig. 4.6e, Ni–$Al_{0.33}O_x$ can be used at least 4 times without apparent deactivation is observed. No Ni and Al ions in filtrate after reaction were detected via

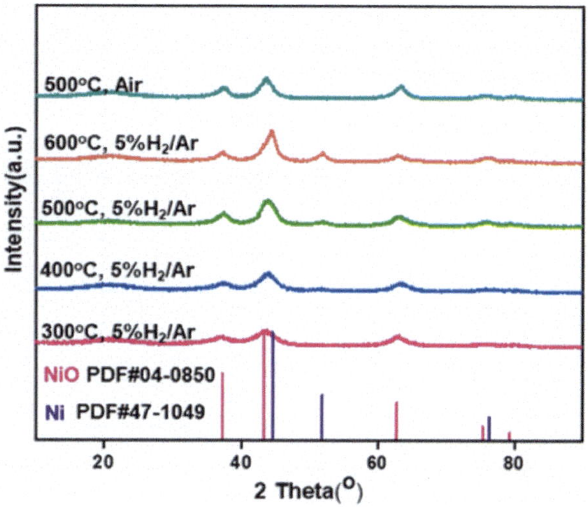

Fig. 4.11 XRD patterns of Ni–$Al_{0.33}O_x$ precursor treated with different reduction temperature

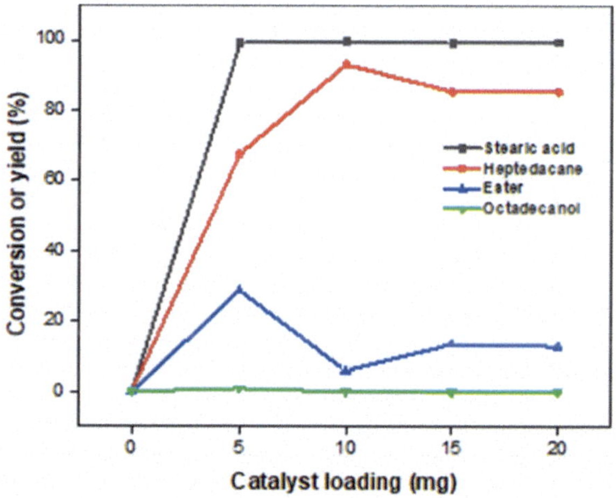

Fig. 4.12 The influence of Ni–$Al_{0.33}O_x$ loading for CTH of stearic acid. Reaction condition: 25 mg stearic acid, 2.5 mL isopropanol, 250 °C, 8 h

ICP-AES test (Table 4.2). The XRD patterns of the fresh and recycled Ni–$Al_{0.33}O_x$ (Run 3) show that diffraction peaks of metallic Ni become sharper after reaction indicating the agglomeration of Ni happened during the HDO process (Fig. 4.6f). The morphology of recycled Ni–$Al_{0.33}O_x$ shown in SEM and TEM images (Figs. 4.16a, b and 4.17) displays a small quantity of agglomerates in the catalyst. However, homogenous distribution of Ni, Al and O observed from EDS line-profile results (Fig. 4.18c, d) implies that Ni and AlO_x remain in good contact. Meanwhile, XPS results show that no visible binding energy shift are observed for Al 2p and Ni 3p between fresh and recycled Ni–$Al_{0.33}O_x$ (Fig. 4.18a), suggesting the charge transfer maintained during the HDO process. Additionally, only a handful decrease of $O_{vacancy}$ content

Fig. 4.13 Gas chromatograms of products distribution for CTH of stearic acid over Ni–$Al_{0.33}O_x$ at 250 °C with different reaction time

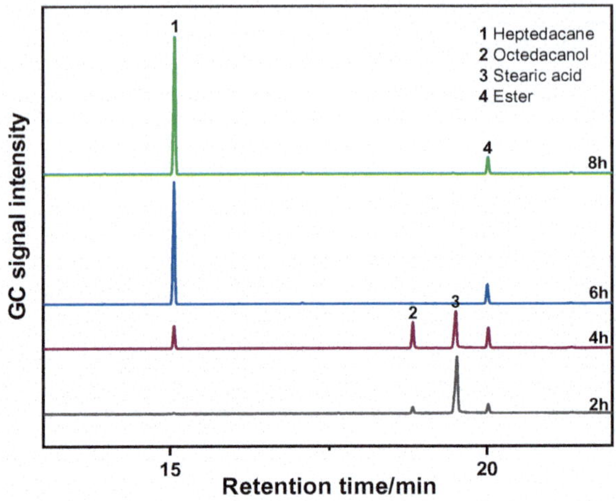

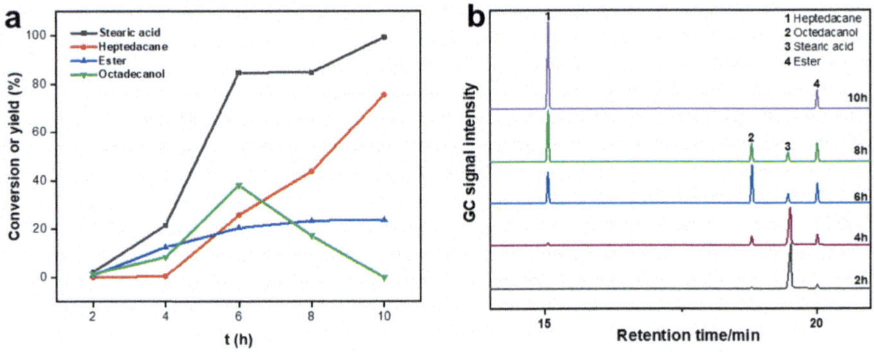

Fig. 4.14 **a** Products distribution and **b** gas chromatograms for CTH of stearic acid over Ni–$Al_{0.33}O_x$ at 230 °C with different reaction time

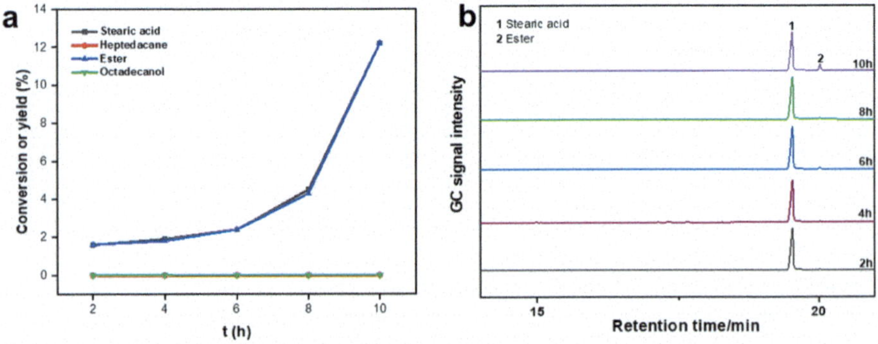

Fig. 4.15 **a** Products distribution and **b** gas chromatograms for CTH of stearic acid over Ni–$Al_{0.33}O_x$ at 200 °C with different reaction time

Table 4.2 ICP results of recycled filtrate after CTH of stearic acid over Ni–$Al_{0.33}O_x$ at 250 °C

	Ni ppm	Al ppm
ICP	0.001	0.0

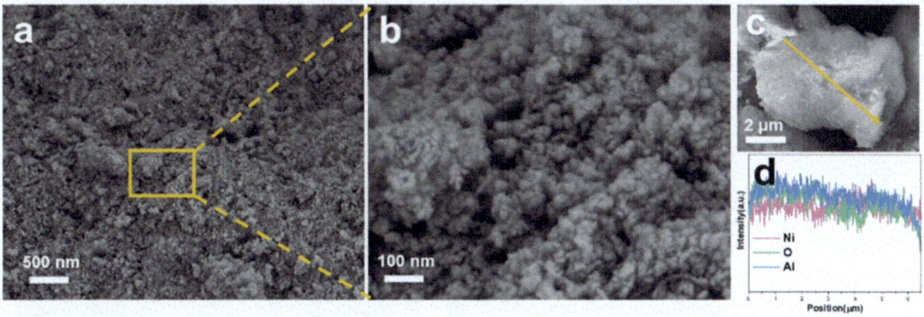

Fig. 4.16 **a** Low and **b** high multiple SEM images of recycled Ni–$Al_{0.33}O_x$; **c, d** EDS line-profile over particles of Ni–$Al_{0.33}O_x$

on the surface of recycled N Ni–$Al_{0.33}O_x$ from O 1 s core-level signal was observed. These results disclose that although the aggregation of metallic Ni nanoparticles would weaken the activity, the presence of adequate OV sites maintains the catalytic performance of Ni–$Al_{0.33}O_x$ to a certain extent.

4.3.1.2 DFT Calculations

In order to verify the role of OV sites and interface between Ni and Al oxides for activating of fatty acid. DFT calculation was conducted and Ni (111) was established containing 5-layer slabs. Al oxides cluster model was selected according by previous reports (Li & Cheng, 2012; Rahane et al., 2011) to represent Al_2O_3 and AlO_x. In Ni–Al_2O_3, a cage-like cluster containing 4 Al atoms and 6 O atom was used to show the properties of solid Al_2O_3. Furthermore, an oxygen atom neighbored to the Ni surface was subtracted from Ni–Al_2O_3 as an initial model of Ni–AlO_x. All model was fully optimized before next calculation. The final state of Ni, Ni–Al_2O_3 and Ni–AlO_x are presented in Fig. 4.19.

The charge density difference was employed to reveal the change in charge spatial distribution. Figure 4.20 shows the electron accumulation and depletion, which are represented by yellow and blue parts, respectively. In the interface region, the yellow area mainly dispensed between Ni and Al atom and the blue area covered the region near the surface Ni–O (Fig. 4.20a, b). It was found that, in the interface region, the electrons were concentrated at the interface of Ni and Al oxide. The xy-planar-averaged differential charge density was calculated and is presented in Fig. 4.20c, d. The intensity of $\Delta\rho$ in Ni–Al_2O_3 and Ni–AlO_x show that the interface charge density

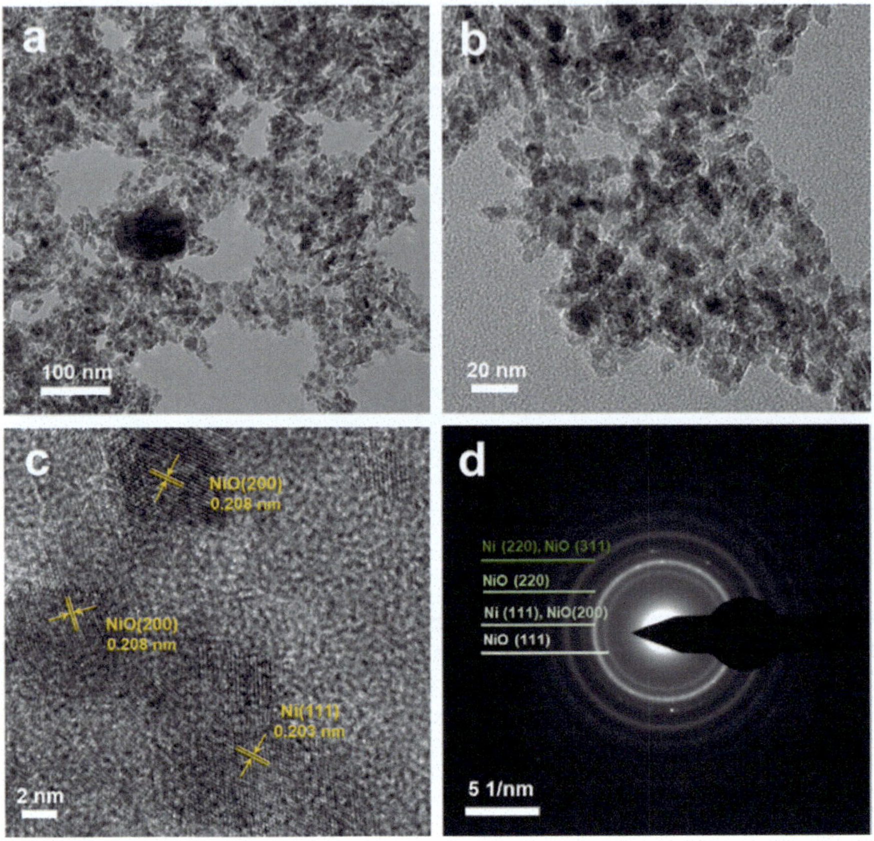

Fig. 4.17 **a** Low and **b** high multiple TEM images and of recycled Ni–$Al_{0.33}O_x$; **c**, **d** HRTEM image, intensity profiles of d-spacing and SAED of recycled Ni–$Al_{0.33}O_x$

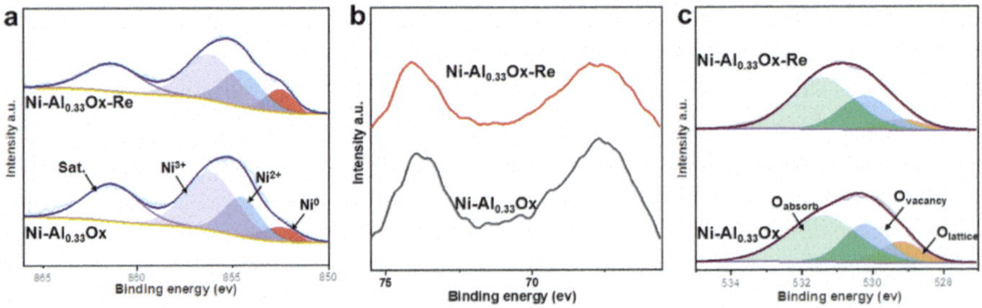

Fig. 4.18 Ni–$Al_{0.33}O_x$ and recycled Ni–$Al_{0.33}O_x$ XPS spectra of **a** Ni $2p_{3/2}$, **b** Al 2p, Ni 3p, **c** O 1 s

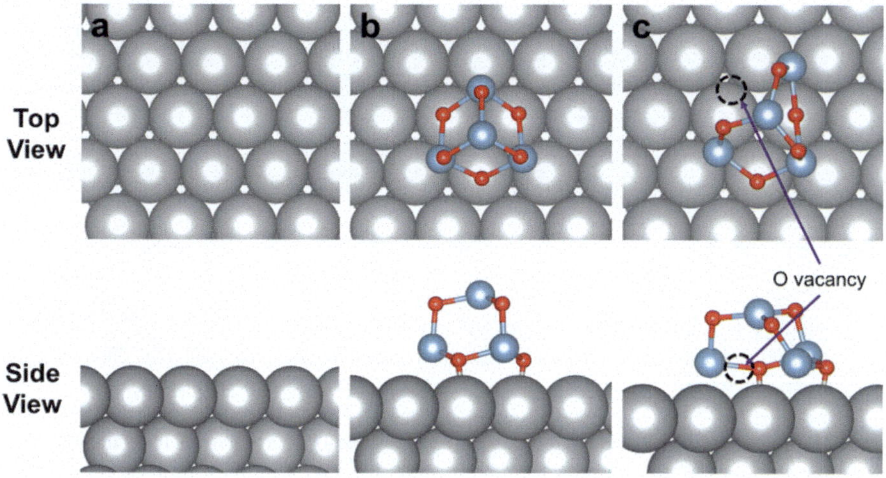

Fig. 4.19 Calculated structure of **a** Ni(111), **b** Ni–Al_2O_3 and **c** Ni–AlO_x. The gray, blue and red balls represent Ni, Al and O atom, respectively

raised up in Ni–AlO_x, implying the charge transfer enhanced after induction of the OV in Al oxide, in good agreement with XPS results.

Absorbing carboxyl group of fatty acid plays an important role for activating fatty acid HDO process (Kordulis et al., 2016; Xing et al., 2020; Zhang et al., 2018a).

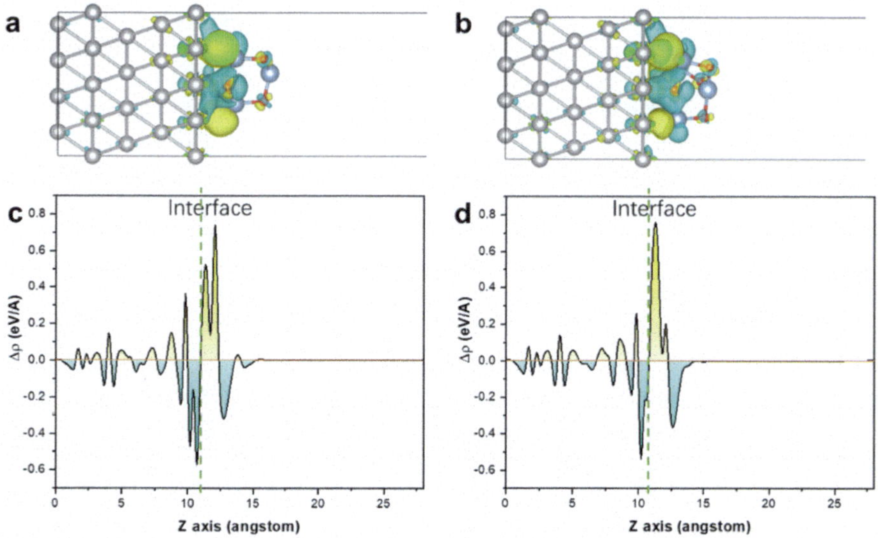

Fig. 4.20 The electron accumulation and depletion of **a**, **c** Ni–Al_2O_3 and **b**, **d** Ni–AlO_x. (Yellow and blue parts represent electron accumulation and depletion, respectively)

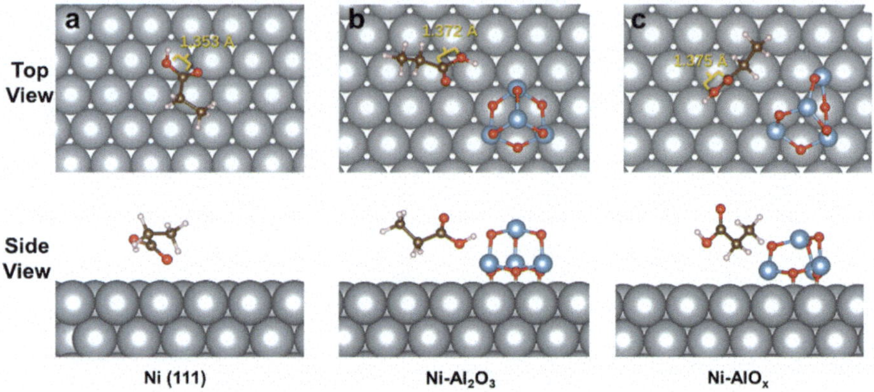

Fig. 4.21 Calculated C–OH bond length and adsorption energy of propionic acid on **a** Ni (111), **b** Ni–Al_2O_3 and **c** Ni–AlO_x

Thus, the propionic acid molecule as a model compound was adsorbed on the Ni (111), Ni–Al_2O_3 and Ni–AlO_x, respectively in Fig. 4.21. The surface with OV will slightly stretch the C–OH bond length from 1.372 to 1.375 Å, which indicated the propionic acid molecule can be activate easier to participate in a reaction. Moreover, the adsorption energy was calculated by follow equation.

$E_{ad} = E_{molecule/surface} - E_{surface} - E_{molecule}$.

The adsorption energy here are exoergic on pure Ni (111) surface, Ni–Al_2O_3 and Ni–AlO_x with −0.778, −0.675 and −1.235 eV, respectively. Note that interface between Ni and Al_2O_3 cannot lower the adsorption energy. Whereas Ni–AlO_x shows lowest adsorption energy indicating that this mode of adsorption on the surface with OV site is more favorable than adsorption of propionic acid on the Ni and Ni–Al_2O_3. This reveals surface OV will result in increased propionic acid adsorption, which facilitates the reaction, in good agreement with the excellent HDO performance of Ni–$Al_{0.33}O_x$.

4.4 Conclusions

In summary, we developed an oxygen-vacancy-rich Ni–$Al_{0.33}O_x$ for in-situ HDO of stearic acid into n-heptadecane using isopropanol as hydrogen donor at mild condition. As compared to Ni powder and Ni/Al_2O_3-IMP, hydrothermal derived Ni–$Al_{0.33}O_x$ possesses smaller particle size and higher amount of oxygen vacancies, which are beneficial to expose the active sites and promote the HDO of stearic acid. Electron exchange between Ni and AlO_x observed by XPS demonstrates the metal/support good interaction. The highest yield of n-heptadecane was 93.2% over Ni–$Al_{0.33}O_x$ at 250 °C for 8 h. Moreover, Ni–$Al_{0.33}O_x$ displayed favorable stability and could be used at least 4 times without loss activity. DFT calculation further demonstrates that oxygen-vacancy-rich Ni–AlO_x interface is more favorable for fatty acids adsorption than oxygen-vacancy-poor Ni/Al_2O_3 interface and pure Ni surface. Enhanced fatty acid adsorption contributes to the activation of carboxyl groups, thus promotes the HDO process of fatty acid.

References

Ali, A., & Zhao, C. (2020). Ru nanoparticles supported on hydrophilic mesoporous carbon catalyzed low-temperature hydrodeoxygenation of microalgae oil to alkanes at aqueous-phase. *Chinese Journal of Catalysis, 41*, 1174–1185.

Cao, X., Long, F., Zhai, Q., Liu, P., Xu, J., & Jiang, J. (2020). Enhancement of fatty acids hydrodeoxygenation selectivity to diesel-range alkanes over the supported Ni–MoO_x catalyst and elucidation of the active phase. *Renewable Energy, 162*, 2113–2125.

Chen, C.-Y., Yeh, K.-L., Aisyah, R., Lee, D.-J., & Chang, J.-S. (2011). Cultivation, photobioreactor design and harvesting of microalgae for biodiesel production: A critical review. *Bioresource Technology, 102*, 71–81.

Chen, H., Wang, Q., Zhang, X., & Wang, L. (2015). Quantitative conversion of triglycerides to hydrocarbons over hierarchical ZSM-5 catalyst. *Applied Catalysis B: Environmental, 166–167*, 327–334.

Chen, L., Zhang, F., Li, G., & Li, X. (2017). Effect of Zn/Al ratio of Ni/ZnO-Al_2O_3 catalysts on the catalytic deoxygenation of oleic acid into alkane. *Applied Catalysis a: General, 529*, 175–184.

Crawford, J. M., & Carreon, M. A. (2018). Decarboxylation of diunsaturated linoleic acid to heptadecane over zeolite supported Pt/ZIF-67 catalysts. *Industrial and Engineering Chemistry Research, 57*, 15991–15997.

Cui, D., Xu, K., Dong, X., Lv, D., Dong, F., Hao, W., Du, Y., & Chen, J. (2020). Controlled hydrogenation into defective interlayer bismuth oxychloride via vacancy engineering. *Communications Chemistry, 3*, 73.

Ding, S., Parlett, C. M. A., & Fan, X. (2022). Recent developments in multifunctional catalysts for fatty acid hydrodeoxygenation as a route towards biofuels. *Molecular Catalysis, 523*, 111492.

Fu, J., Yang, C., Wu, J., Zhuang, J., Hou, Z., & Lu, X. (2015). Direct production of aviation fuels from microalgae lipids in water. *Fuel, 139*, 678–683.

He, L., Huang, Y., Wang, A., Wang, X., Chen, X., Delgado, J. J., & Zhang, T. (2012). A noble-metal-free catalyst derived from Ni–Al hydrotalcite for hydrogen generation from $N_2H_4 \cdot H_2O$ decomposition. *Angewandte Chemie International Edition, 51*, 6191–6194.

Jia, J., Qian, C., Dong, Y., Li, Y. F., Wang, H., Ghoussoub, M., Butler, K. T., Walsh, A., & Ozin, G. A. (2017). Heterogeneous catalytic hydrogenation of CO_2 by metal oxides: Defect engineering—Perfecting imperfection. *Chemical Society Reviews, 46*, 4631–4644.

Jiraroj, D., Jirarattanapochai, O., Anutrasakda, W., Samec, J. S. M., & Tungasmita, D. N. (2021). Selective decarboxylation of biobased fatty acids using a Ni–FSM-16 catalyst. *Applied Catalysis B: Environmental, 291*, 120050.

Kon, K., Onodera, W., Takakusagi, S., & Shimizu, K.-I. (2014). Hydrodeoxygenation of fatty acids and triglycerides by Pt-loaded Nb_2O_5 catalysts. *Catalysis Science & Technology, 4*, 3705–3712.

Kong, X., Fang, Z., Bao, X., Wang, Z., Mao, S., & Wang, Y. (2018). Efficient hydrogenation of stearic acid over carbon coated NiFe catalyst. *Journal of Catalysis, 367*, 139–149.

Kordulis, C., Bourikas, K., Gousi, M., Kordouli, E., & Lycourghiotis, A. (2016). Development of nickel based catalysts for the transformation of natural triglycerides and related compounds into green diesel: A critical review. *Applied Catalysis B: Environmental, 181*, 156–196.

Li, R., & Cheng, L. (2012). Structural determination of $(Al_2O_3)_n$ (n=1–7) clusters based on density functional calculation. *Computational & Theoretical Chemistry, 996*, 125–131.

Li, W., Gao, Y., Yao, S., Ma, D., & Yan, N. (2015). Effective deoxygenation of fatty acids over $Ni(OAc)_2$ in the absence of H_2 and solvent. *Green Chemistry, 17*, 4198–4205.

Li, M., Fu, J., Xing, S., Yang, L., Zhang, X., Lv, P., Wang, Z., & Yuan, Z. (2020). A novel catalyst with variable active sites for the direct hydrogenation of waste oils into jet fuel. *Applied Catalysis B: Environmental, 260*, 118114.

Li, Z., Lu, X., Sun, W., Leng, L., Zhang, M., Li, H., Bai, L., Yuan, D., Horton, J. H., Xu, Q., & Wang, J. (2021). One-step synthesis of single palladium atoms in $WO_{2.72}$ with high efficiency in chemoselective hydrodeoxygenation of vanillin. *Applied Catalysis B: Environmental, 298*, 120535.

Liu, H., Jia, W., Yu, X., Tang, X., Zeng, X., Sun, Y., Lei, T., Fang, H., Li, T., & Lin, L. (2021). Vitamin C-assisted synthesized Mn–Co oxides with improved oxygen vacancy concentration: Boosting lattice oxygen activity for the air-oxidation of 5-(Hydroxymethyl)furfural. *ACS Catalysis, 11*, 7828–7844.

Lu, Y., Zhang, Z., Wang, H., & Wang, Y. (2021). Toward efficient single-atom catalysts for renewable fuels and chemicals production from biomass and CO_2. *Applied Catalysis B: Environmental, 292*, 120162.

Luo, J., Dong, Y., Petit, C., & Liang, C. (2021). Development of gold catalysts supported by unreducible materials: Design and promotions. *Chinese Journal of Catalysis, 42*, 670–693.

Lycourghiotis, S., Kordouli, E., Sygellou, L., Bourikas, K., & Kordulis, C. (2019). Nickel catalysts supported on palygorskite for transformation of waste cooking oils into green diesel. *Applied Catalysis B: Environmental, 259*, 118059.

Meng, C., Zhao, G., Shi, X.-R., Chen, P., Liu, Y., & Lu, Y. (2021). Oxygen-deficient metal oxides supported nano-intermetallic $InNi_3C_{0.5}$ toward efficient CO_2 hydrogenation to methanol. *Science Advances, 7*, eabi6012.

Miao, C., Marin-Flores, O., Davidson, S. D., Li, T., Dong, T., Gao, D., Wang, Y., Garcia-Pérez, M., & Chen, S. (2016). Hydrothermal catalytic deoxygenation of palmitic acid over nickel catalyst. *Fuel, 166*, 302–308.

Ni, J., Leng, W., Mao, J., Wang, J., Lin, J., Jiang, D., & Li, X. (2019). Tuning electron density of metal nickel by support defects in Ni/ZrO_2 for selective hydrogenation of fatty acids to alkanes and alcohols. *Applied Catalysis B: Environmental, 253*, 170–178.

Rahane, A. B., Deshpande, M. D., & Kumar, V. (2011). Structural and electronic properties of $(Al_2O_3)_n$ Clusters with $n = 1$–10 from first principles calculations. *Journal of Physical Chemistry C, 115*, 18111–18121.

Rakmae, S., Osakoo, N., Pimsuta, M., Deekamwong, K., Keawkumay, C., Butburee, T., Faungnawakij, K., Geantet, C., Prayoonpokarach, S., Wittayakun, J., & Khemthong, P. (2020). Defining nickel phosphides supported on sodium mordenite for hydrodeoxygenation of palm oil. *Fuel Processing Technology, 198*, 106236.

Shim, J.-O., Jang, W.-J., Jeon, K.-W., Lee, D.-W., Na, H.-S., Kim, H.-M., Lee, Y.-L., Yoo, S.-Y., Jeon, B.-H., Roh, H.-S., & Ko, C. H. (2018). Petroleum like biodiesel production by catalytic decarboxylation of oleic acid over $Pd/Ce\text{-}ZrO_2$ under solvent-free condition. *Applied Catalysis a: General, 563*, 163–169.

Sun, K., Wilson, A. R., Thompson, S. T., & Lamb, H. H. (2015). Catalytic deoxygenation of octanoic acid over supported palladium: Effects of particle size and alloying with gold. *ACS Catalysis, 5*, 1939–1948.

Wang, J., Nie, R., Xu, L., Lyu, X., & Lu, X. (2019). Catalytic transfer hydrogenation of oleic acid to octadecanol over magnetic recoverable cobalt catalysts. *Green Chemistry, 21*, 314–320.

Wang, J., Xu, L., Nie, R., Lyu, X., & Lu, X. (2020a). Bifunctional $CuNi/CoO_x$ catalyst for mild-temperature in situ hydrodeoxygenation of fatty acids to alkanes using isopropanol as hydrogen source. *Fuel, 265*, 116913.

Wang, L., Niu, X., & Chen, J. (2020b). SiO_2 supported Ni–In intermetallic compounds: Efficient for selective hydrogenation of fatty acid methyl esters to fatty alcohols. *Applied Catalysis B: Environmental, 278*, 119293.

Wu, J., Shi, J., Fu, J., Leidl, J. A., Hou, Z., & Lu, X. (2016). Catalytic decarboxylation of fatty acids to aviation fuels over Nickel supported on activated carbon. *Science Rep-UK, 6*, 27820.

Xin, H., Zhou, W., Zhou, K., Du, X., Li, D., & Hu, C. (2019). Controlling the growth of activated carbon supported nickel phosphide catalysts via adjustment of surface group distribution for hydrodeoxygenation of palmitic acid. *Catalysis Today, 319*, 182–190.

Xing, S., Liu, Y., Liu, X., Li, M., Fu, J., Liu, P., Lv, P., & Wang, Z. (2020). Solvent-free hydrodeoxygenation of bio-lipids into renewable alkanes over NiW bimetallic catalyst under mild conditions. *Applied Catalysis B: Environmental, 269*, 118718.

Yang, C., Nie, R., Fu, J., Hou, Z., & Lu, X. (2013). Production of aviation fuel via catalytic hydrothermal decarboxylation of fatty acids in microalgae oil. *Bioresource Technology, 146*, 569–573.

Yao, X., Strathmann, T. J., Li, Y., Cronmiller, L. E., Ma, H., & Zhang, J. (2021). Catalytic hydrothermal deoxygenation of lipids and fatty acids to diesel-like hydrocarbons: A review. *Green Chemistry, 23*, 1114–1129.

Yeh, T. M., Hockstad, R. L., Linic, S., & Savage, P. E. (2015). Hydrothermal decarboxylation of unsaturated fatty acids over $PtSn_x/C$ catalysts. *Fuel, 156*, 219–224.

Zhang, Z., Chen, H., Wang, C., Chen, K., Lu, X., Ouyang, P., & Fu, J. (2018a). Efficient and stable $Cu\text{-}Ni/ZrO_2$ catalysts for in situ hydrogenation and deoxygenation of oleic acid into heptadecane using methanol as a hydrogen donor. *Fuel, 230*, 211–217.

Zhang, Z., Yang, Q., Chen, H., Chen, K., Lu, X., Ouyang, P., Fu, J., & Chen, J. G. (2018b). In situ hydrogenation and decarboxylation of oleic acid into heptadecane over a Cu–Ni alloy catalyst using methanol as a hydrogen carrier. *Green Chemistry, 20*, 197–205.

Zhang, Z., Bi, G., Zhang, H., Zhang, A., Li, X., & Xie, J. (2019). Highly active and selective hydrodeoxygenation of oleic acid to second generation bio-diesel over SiO_2-supported $Co_xNi_{1-x}P$ catalysts. *Fuel, 247*, 26–35.

Zheng, Y., Wang, J., Liu, C., Lu, Y., Lin, X., Li, W., & Zheng, Z. (2020). Efficient and stable Ni–Cu catalysts for ex situ catalytic pyrolysis vapor upgrading of oleic acid into hydrocarbon: Effect of catalyst support, process parameters and Ni-to-Cu mixed ratio. *Renewable Energy, 154*, 797–812.

Zhong, J., Deng, Q., Cai, T., Li, X., Gao, R., Wang, J., Zeng, Z., Dai, G., & Deng, S. (2021). Graphitic carbon embedded FeNi nanoparticles for efficient deoxygenation of stearic acid without using hydrogen and solvent. *Fuel, 292*, 120248.

Zhou, Y., Liu, L., Li, G., & Hu, C. (2021). Insights into the influence of ZrO_2 crystal structures on methyl laurate hydrogenation over Co/ZrO_2 catalysts. *ACS Catalysis, 11*, 7099–7113.

Part III
Water as Hydrogen Source

Chapter 5
Chemoselective Synthesis of Propionic Acid from Biomass and Lactic Acid Over a Cobalt Catalyst in Aqueous Media

5.1 Introduction

Propionic acid (PA) is a good antifungal agents in food industry and also has been extensively applied as a crucial chemical intermediate in industrial manufacture. Such as in the synthesis of cellulose fibers, perfumes, herbicides and pharmaceuticals, artificial fruit flavors, and preservatives for food, animal feed, and grain, etc. (Boyaval & Corre, 1995; Hsu & Yang, 1991; Martínez-Campos & de la Torre, 2002). Due to its excellent industrial application value, the synthesis of PA has attracted much attention. Nowadays, PA production is strongly done by petrochemical routes, such as the direct synthesis of carboxide and ethylene, oxidation of propionaldehyde and by-product of acetic synthesis by light paraffin oxidation. However, the feasibility of PA production based on the petrochemical industry is facing serious challenges due to the emergence of environment pollution and energy shortage.

Recently, the utilization of biomass and derived materials to synthesize chemical stocks has been proved a promising method to replace the petroleum-derived materials since it is renewable, low-cost, less pollution and abundant (Bond et al., 2014, Corma et al., 2007, Dauenhauer & Huber, 2014; Gallezot, 2012, Huber et al., 2006, Schmidt & Dauenhauer, 2007). Considerable efforts have been made to effectively convert biomass resource into PA (Dietz et al., 2013; Dishisha et al., 2015; Feng et al., 2011; Jiang et al., 2015; Lewis & Yang, 1992; Liu et al., 2011; Martínez-Campos & de la Torre, 2002). However, to the best of our knowledge, only fermentative processes for PA production have been reported in the last decades. Majority of studies on PA fermentation production accomplished through Propionibacterium acidipropionici have focused on the conversion of glucose and whey lactose (Dietz et al., 2013; Jiang et al., 2015; Martínez-Campos & de la Torre, 2002). Also, some studies have employed glycerol or lactic acid (LA) as a carbon source on account that less acetic acid was produced during consumption of glycerol/LA (Dishisha et al., 2015; Liu et al., 2011). In terms of the properties of fermentation process: unstable product quality, undesirable salt waste effluents, small space–time yields and high purification cost, the fermentation process lacks economically competitive ability in

Z. Huo, *Diverse Hydrogen Sources for Biomass-derivatives Conversion*,
https://doi.org/10.1007/978-981-99-1673-3_5

industrial application in the long run. Thus, it is extremely essential to develop more efficient methods to product PA.

The production of value-added chemicals, such as formic acid, acetic acid, LA, levulinic acid and alcohols, etc., based on the chemical conversion of the biomass resource has been reported recently (Chen et al., 2014; Ding et al., 2014; Gao et al., 2013, 2016; Jin & Enomoto, 2009, 2011; Miao et al., 2015; Ren et al., 2016a, 2016b; Testova et al., 2014; Wang et al., 2013, 2015; Xu et al., 2014). However, no any methods for PA production from the biomass conversion was reported because the degradation products of big molecule biomass are always low-carbon organics (carbon numbers less than 3). Even if big molecule biomass can be effectively degraded into C_3 compounds under specific reaction conditions, but the ultimate products are always LA, propylene glycol or glycerol but not PA. Hence, the production of PA by the redox of C_3 compounds may have a promising future. Previously, we unexpectedly found the small amount of PA, as another reduction product of LA, was always detected during the process for the selective conversion of LA into 1,2-propylene glycol (Xiao et al., 2015), which offered the possibility of PA production from LA.

Theoretically, LA can be selectively reduced into 1,2-propylene glycol or PA under proper reaction conditions (Fig. 5.1). However, hydroxyl (-OH) is difficultly reduced in that the oxidability of carboxyl (-COOH) is much higher than hydroxyl group, for which the enthalpy of reaction (ΔH_r) estimated by bonding energy can partly account for the reactivity of the hydroxyl and carboxyl group (Eqs. 5.1 and 5.2). Therefore, the investigation of redox of hydroxyl in the presence of carboxyl may have a signification implication for the organic technology.

$$CH_3CH(OH)COOH + H_2 \underset{\Delta H_r(1)=-118\,kj/mol}{\longrightarrow} CH_3CH_2COOH + H_2O \quad (5.1)$$

$$CH_3CH(OH)COOH + 2H_2 \underset{\Delta H_r(2)=-85\,kj/mol}{\longrightarrow} CH_3CH(OH)CH_2OH + H_2O \quad (5.2)$$

The aim of this study is to develop a new process for PA production from the biomass-derived LA and carbohydrates in high temperature water. Fermentation production of PA from biomass-derived materials has been thoroughly investigated in the past several years. However, the research about the chemical process for PA production by utilizing biomass as the raw materials is still in its infancy. Here, we first reported that hydrothermal method is applied to bring off the chemical process for the production of PA from LA and carbohydrates (Fig. 5.2).

$$CH_3CH(OH)COOH + H_2 \longrightarrow CH_3CH_2COOH + H_2O \quad (1)$$
$$\Delta H_r(1) = -118 \text{ kJ/mol}$$
$$CH_3CH(OH)COOH + 2H_2 \longrightarrow CH_3CH(OH)CH_2OH + H_2O \quad (2)$$
$$\Delta H_r(2) = -85 \text{ kJ/mol}$$

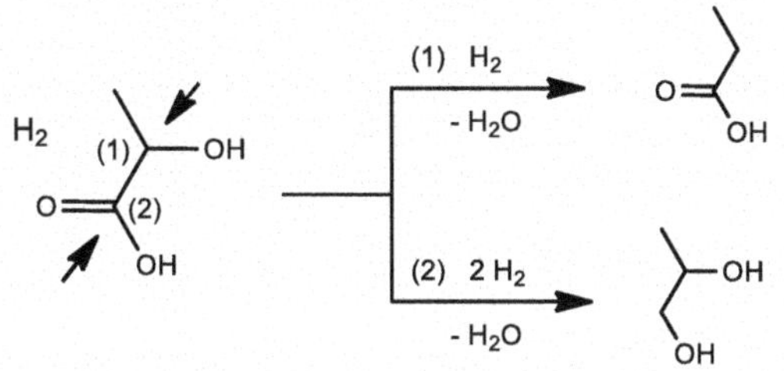

Fig. 5.1 Reprinted from Royal Society of Chemistry (Ren et al., 2017)

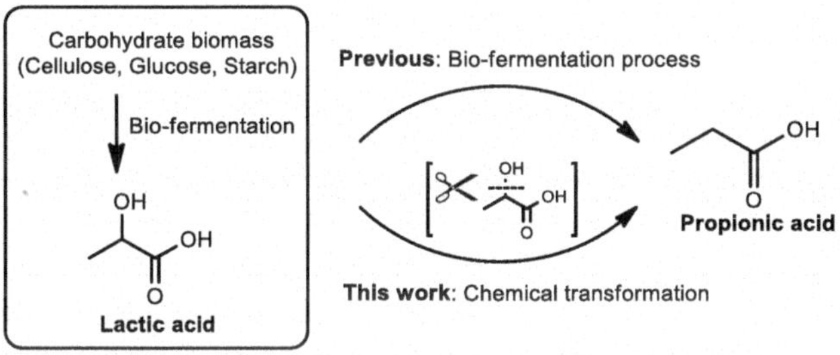

Fig. 5.2 PA production from carbohydrates and LA over Co catalyst. Reprinted from Royal Society of Chemistry (Ren et al., 2017)

5.2 Materials and Methods

5.2.1 Materials

LA (DL, ~ 90%T) was obtained from Sigma-Aldrich Chemical Reagent Co., Ltd. Zn powder (AR) was obtained from Aladdin Chemical Reagent Co. Ltd. Co powder (200-mesh, 2N), other metals and metal oxides were obtained from Sinopharm Chemical Reagent Co. Ltd., China. The standard reagents PA (Sigma- Aldrich, > 99.7%), 1,2-propylene glycol (PG) (J&K Chemical, > 99.8%), Hydroxyacetone (HA) (Alfa Aesar, > 95%), acetic acid (AA) (Sigma-Aldrich, > 99.7%) and 1.0 N Standardized LA (Alfa Aesar) were employed in qualitative analysis of liquid example.

5.2.2 Experimental Procedure

Experiments were performed by using a series of batch Teflon reactors (inner volume: 30 mL). The procedure was conducted as follows. Firstly, the desired concentration of LA solution was prepared beforehand according to the desired water volume. Then, the desired catalyst, reductant and LA solution were added into the reactor chamber. Reactor was brought into the preheated oven for 20 min to rise to the desired reaction temperature. After preset reaction time, the reactor was removed from the oven and cooled down to the room temperature with the help of electric fan. Finally, liquid sample was collected and filtered through a 0.45 μm syringe for analysis. Solid sample was collected and washed with deionized water and ethanol several times and dried in the oven at 50 °C for 24 h.

5.2.3 Products Analysis

The liquid samples were analyzed by gas chromatography/mass spectroscopy (GC–MS), total organic carbon (TOC), high-performance liquid chromatography (HPLC). The Agilent 7890 GC/MS system was equipped with 5985C inert mass selective detector (MSD) with triple-axis detector while the Agilent 1200 HPLC system was equipped with Variable wavelength detector (VW) and Refractive index detector (RI) with a 2 mmol/L H_2SO_4 solution as the mobile phase at a flow rate of 1.0 mL/min. The Liquid sample was also tested by inductively coupled plasma-optical emission spectrometry (ICP-OES) to measure the metal ion concentration. Solid sample washed with deionized water and ethanol several times was dried in an oven at 50 °C for 24 h. Then the precipitate was analyzed by XRD (Bruker D8 Advance X-Ray Diffractometer, equipped with Cu Kα radiation: $\lambda = 1.5406$ Å, scanning rate: 0.02° s^{-1}, 2θ ranges: 10–80°).

The LA conversion ($LA_{Conv.}$), product i yield (Y_i) and selectivity to product i (S_i) were calculated as following equations:

$$LA_{conv.}(\%) = \left(1 - \frac{mole\,of\,lactic\,acid\,at\,particulartime}{initial\,mole\,of\,lactic\,acid}\right) \times 100 \tag{5.3}$$

$$Y_i(\%) = \left(\frac{mole\,of\,product\,i}{initial\,mole\,of\,lactic\,acid}\right) \times 100 \tag{5.4}$$

$$S_i(\%) = \left(\frac{mole\,of\,product\,i}{mole\,of\,all\,product\,formed}\right) \times 100 \tag{5.5}$$

5.3 Results and Discussion

5.3.1 *Catalyst Screening*

Previous research found that small amount of PA as the by-product was always detected during the production of 1,2-propylene glycol from LA using CuO as the catalyst (Xiao et al., 2015). According to the reaction mechanism, PA can be obtained by the selective reduction of hydroxyl of LA. To improve the selectivity of PA from LA, a series of experiments with different catalysts were performed. The results in Table 5.1 demonstrated that Co showed the best catalytic activity with a 30.9% yield of PA with other compounds such as PG, HA, AA and NPA(1-propanol) among all catalyst we investigated (entry 3, also see Fig. 5.3). XRD patterns of solid residues showed in Fig. 5.4 that Co still existed while Zn was oxidized into ZnO, it was obvious that Co acted as a catalyst and Zn acted as a reductant in this process. Further, the concentration of Co ion in liquid sample was then tested by ICP and very small amount of Co ion (112 ppm) was found. Besides, the PA yield was significantly lower in the absence of catalyst and/or reductant, which indicated that both catalyst and reductant were necessary (entries 1 and 2). Other catalysts, such as Cu, CuO, Cu_2O, Ni, Ni_2O_3, Fe_2O_3, Mo, Ti, ZrO_2, SiO_2, TiO_2, MnO_2, Cr, Al_2O_3, were tested and proved to be completely ineffective for the transformation (entries 4–17). It should be noted that the selectivity of PA is 100% when TiO_2 was used as a catalyst, which was much higher than Co catalyst. However, the conversion of LA is very low, only 6% of LA was converted (entry 14).

Co loading had a significant influence in the conversion of LA to PA as shown in Fig. 5.5 and Table 5.2. The yield of PA reached the maximum value of 58.8% with 88.3% LA conversion rate when Co loading was 4 mmol, while yields of PG and NPA showed a slight change with the increase of Co loading from 0 to 10 mmol.

5.3.2 *Effect of Reductant on the Yield of PA*

Next, four active metals, Zn, Al, Fe and Mn, were screened to search for the suitable reductant for the PA production. As shown in Fig. 5.6 and Table 5.3, it was obvious that Zn and Al were effective and gave much higher yields compared to Fe and Mn powders. Zn was chosen as the suitable reductant for the further reaction. From XRD patterns in Figs. 5.7, 5.8 and 5.9, it is worthwhile to note that hydrogen production plays an important role for the LA conversion, the incomplete oxidation of Al, Fe and Mn in water decrease hydrogen production which leads to the low yield of PA.

From Fig. 5.10 and Table 5.4, changing the amount of Zn found that the LA conversion increased from 0 to 20 mmol, and then reached a plateau (96.8%) when Zn amount was increased to 25 mmol. However, PA yield significantly increased with the increase of Zn amount from 5 to 10 mmol, and maximum value of 58.8% yield was attained. A dramatic decrease of PA yield was then observed when Zn

Table 5.1 Catalyst screening for the formation of PA[a]

Entry	Catalysts	Reductant	LA conv. (%)	Yield (%)[b]					Total (%)
				PA	PG	HA	AA	NPA	
1	Co	–	12.1	5.9	0	0	5.5	0	11.4
2	–	Zn	2.3	0	0	0	1.3	0	1.3
3	Co	Zn	96.8	30.9	47.3	< 1.0	1.0	17.8	98.0
4	Cu	Zn	46.0	2.0	40.0	4.0	< 1.0	0	47.0
5	CuO	Zn	66.6	2.8	63.8	< 1.0	< 1.0	0	68.6
6	Cu_2O	Zn	38.3	1.5	36.8	< 1.0	< 1.0	0	40.3
7	Ni	Zn	14.7	< 1.0	14.6	< 1.0	< 1.0	0	17.6
8	Ni_2O_3	Zn	15.3	2.9	10.7	< 1.0	1.7	0	16.3
9	Fe_2O_3	Zn	7.2	3.1	3.2	1.0	< 1.0	0	8.3
10	Mo	Zn	1.0	< 1.0	0	0	< 1.0	0	2.0
11	Ti	Zn	4.3	0	3.1	1.2	0	0	4.3
12	ZrO_2	Zn	1.0	0	0	0	1.0	0	1.0
13	SiO_2	Zn	1.0	0	0	1.0	< 1.0	0	2.0
14	TiO_2	Zn	6.0	6.0	< 1.0	0	< 1.0	0	8.0
15	MnO_2	Zn	1.0	< 1.0	0	0	0	1.0	2.0
16	Cr	Zn	2.0	0	0	0	0	2.0	2.0
17	Al_2O_3	Zn	1.0	< 1.0	0	0	0	1.0	2.0

[a] Reaction conditions: LA 1.5 mmol, catalyst 4 mmol, reductant 25 mmol, H_2O 7.5 mL, 250 °C, 2 h. [b] The yield is calculated as the percentage of products to the initial LA on the carbon base. Reprinted from Royal Society of Chemistry (Ren et al., 2017)

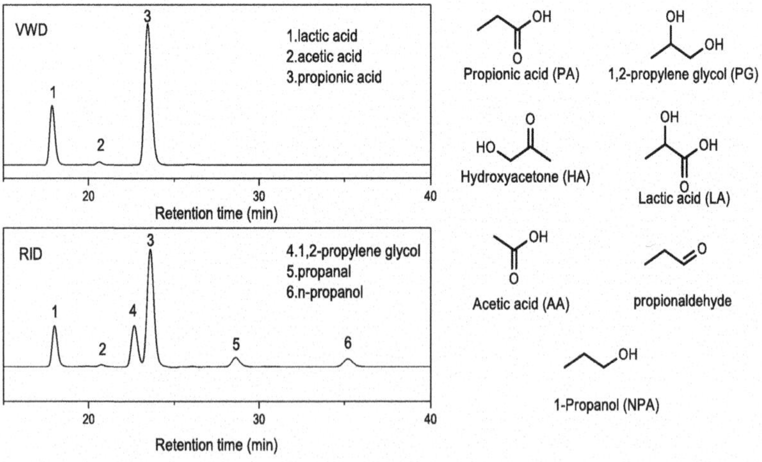

Fig. 5.3 HPLC chromatogram with VW and RI detectors of the liquid sample after reaction (Co 5 mmol, Zn 10 mmol, H_2O 7.5 mL, 250 °C, 2 h). Reprinted from Royal Society of Chemistry (Ren et al., 2017)

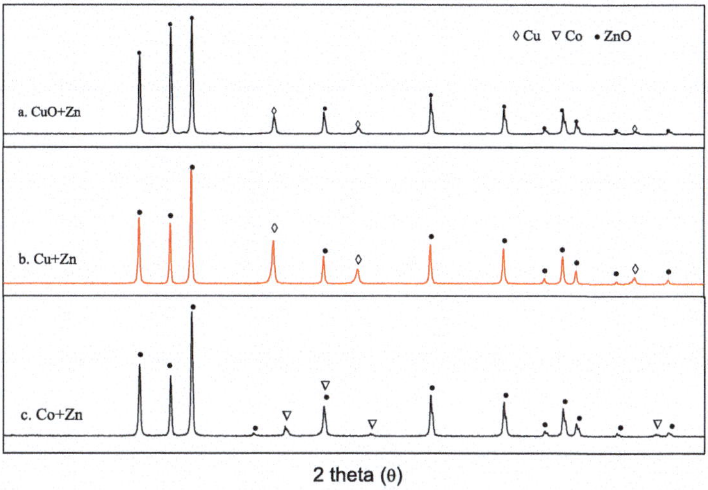

Fig. 5.4 XRD patterns of solid residue after reaction (catalyst 5 mmol, Zn 25 mmol, H_2O 7.5 mL, 250 °C, 2 h). Reprinted from Royal Society of Chemistry (Ren et al., 2017)

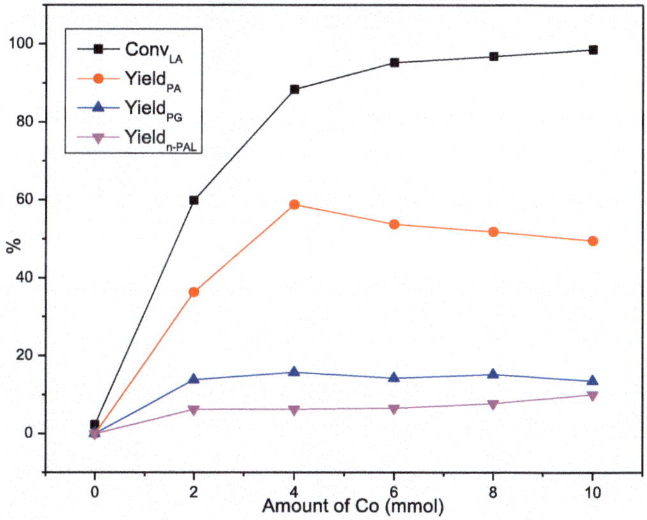

Fig. 5.5 Effect of Co loading (LA 1.5 mmol, Zn 10 mmol, H_2O 7.5 mL, 250 °C, 2 h). Reprinted from Royal Society of Chemistry (Ren et al., 2017)

amount increased after 10 mmol while the yields of PG and NPA further increased, which indicated that more Zn could facilitate the selectivity of LA into PG and the hydrogenation of PA into NPA.

Table 5.2 Effect of Co loading

Entry	Co loading (mmol)	LA conv. (%)	Yield (%)			Total
			PA	PG	NPA	
1	0	2.3	0	0	0	0
2	2	59.8	36.4	13.9	6.3	56.6
3	4	88.3	58.8	15.8	6.3	80.9
4	6	95.2	53.8	14.3	6.6	74.7
5	8	96.8	51.9	15.2	7.8	74.9
6	10	98.6	49.7	13.6	10.2	73.5

Reaction conditions: LA 1.5 mmol, Zn 10 mmol, H_2O 7.5 mL, 250 °C, 2 h. Reprinted from Royal Society of Chemistry (Ren et al., 2017)

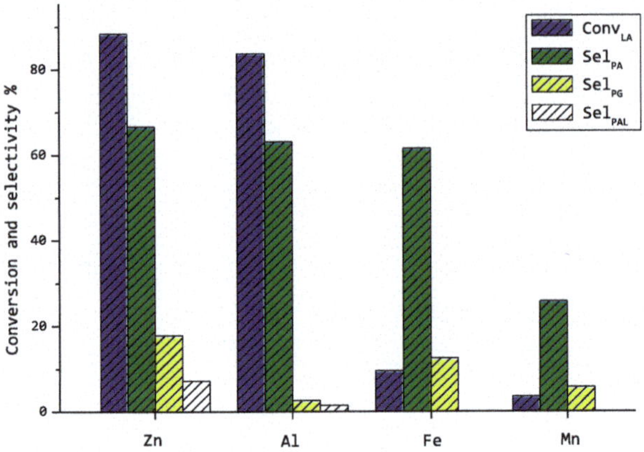

Fig. 5.6 Effect of reductants (LA 1.5 mmol, reductant 10 mmol, Co 4 mmol, H_2O 7.5 mL, 250 °C, 2 h). Reprinted from Royal Society of Chemistry (Ren et al., 2017)

Table 5.3 Effect of reductants[a]

Entry	Reductant	LA conv. (%)	Selectivity (%)			Total yield (%)
			PA	PG	NPA	
1	Zn	88.3	66.6	17.9	7.1	80.9
2[b]	Al	83.7	63.0	2.6	1.4	56.2
3	Fe	9.6	61.4	12.5	0	7.1
4	Mn	3.5	25.7	5.7	0	1.1

[a] Reaction conditions: LA 1.5 mmol, reductant 10 mmol, Co 4 mmol, H_2O 7.5 mL, 250 °C, 2 h.
[b] Part of Al was oxidized, but oxides was not detected by XRD. Reprinted from Royal Society of Chemistry (Ren et al., 2017)

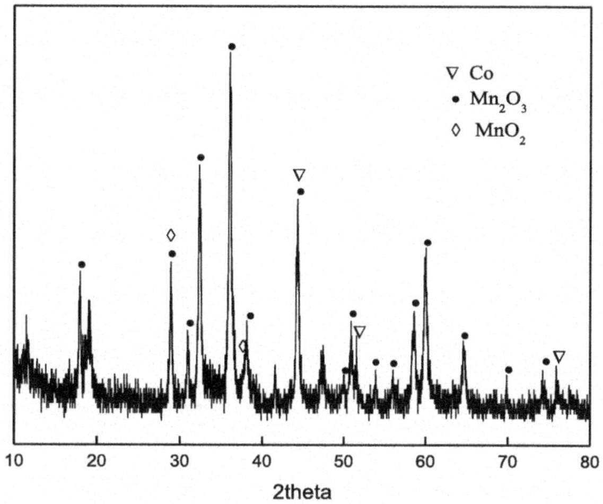

Fig. 5.7 XRD patterns of precipitates with Mn (Co 5 mmol, Mn 10 mmol, H_2O 7.5 mL, 250 °C, 2 h). Reprinted from Royal Society of Chemistry (Ren et al., 2017)

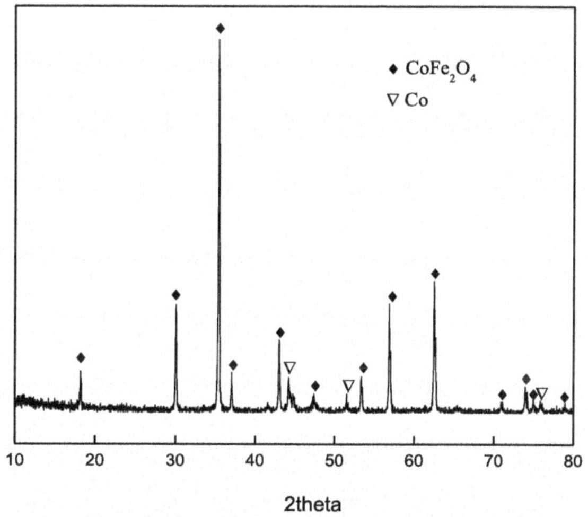

Fig. 5.8 XRD patterns of precipitates with Fe (Co 5 mmol, Fe 10 mmol, H_2O 7.5 mL, 250 °C, 2 h). Reprinted from Royal Society of Chemistry (Ren et al., 2017)

5.3.3 *Influence of Parameters on the Conversion of LA*

Various parameters, such as water volume, temperature and reaction time were also investigated. As illustrated in Fig. 5.11 and Table 5.5, with the increase of water volume from 6 to 13.5 mL, a decline of LA conversion was observed while yield of PA wavered within a narrow range, the yield of PG decreased a little and yield of NPA kept almost constant. The results showed that the decrease of concentration of LA due to the high water volume had a negative effect on the conversion reaction. In view of the pressure, a lower water volume of 7.5 mL with good PA yield was chosen for the following study.

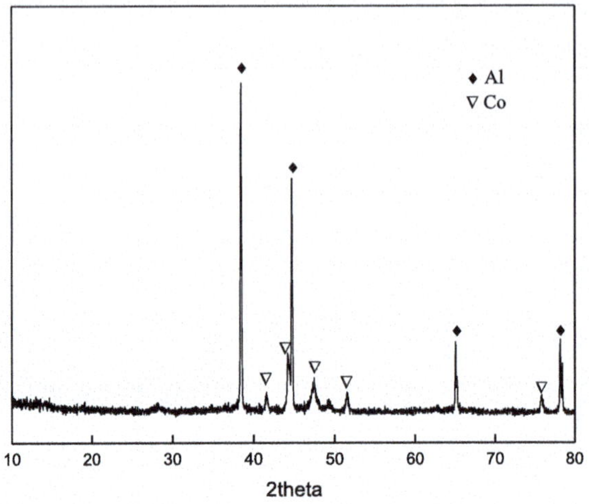

Fig. 5.9 XRD patterns of precipitates with Al (Co 5 mmol, Al 10 mmol, H_2O 7.5 mL, 250 °C, 2 h). Reprinted from Royal Society of Chemistry (Ren et al., 2017)

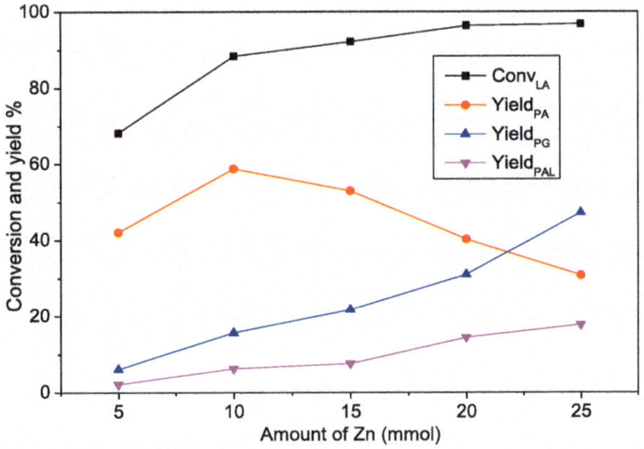

Fig. 5.10 Effect of Zn amount (LA 1.5 mmol, Co 4 mmol, H_2O 7.5 mL, 250 °C, 2 h). Reprinted from Royal Society of Chemistry (Ren et al., 2017)

Table 5.4 Effect of amount of Zn

Entry	Amount (mmol)	LA conv. (%)	Yield (%)			Total
			PA	PG	NPA	
1	5	68.2	42.1	6.1	2.2	50.4
2	10	88.3	58.8	15.8	6.3	80.9
3	15	92.1	53.0	21.8	7.6	82.4
4	20	96.4	40.3	31.1	14.5	85.9
5	25	96.8	30.9	47.3	17.8	96.0

Reaction conditions: LA 1.5 mmol, Co 4 mmol, H_2O 7.5 mL, 250 °C, 2 h. Reprinted from Royal Society of Chemistry(Ren et al., 2017)

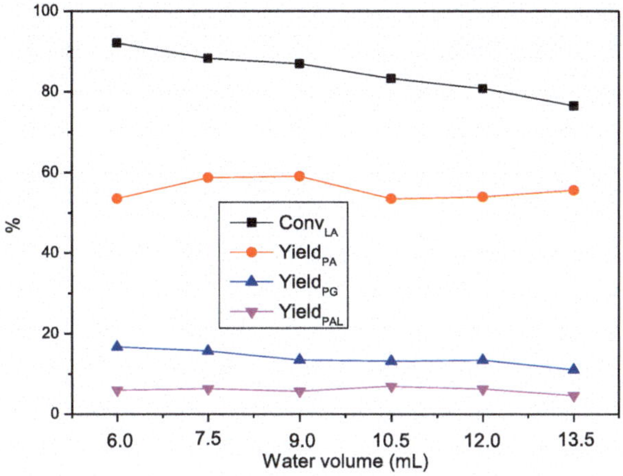

Fig. 5.11 Effect of water volume (LA 1.5 mmol, Zn 10 mmol, Co 4 mmol, 250 °C, 2 h). Reprinted from Royal Society of Chemistry(Ren et al., 2017)

Table 5.5 Effect of water volume

Entry	Water volume (mL)	LA conv. (%)	Yield (%)			Total
			PA	PG	NPA	
1	6.0	92.1	53.6	16.7	6.0	76.3
2	7.5	88.3	58.8	15.8	6.3	80.9
3	9.0	86.9	59.1	13.5	5.7	78.3
4	10.5	83.3	53.6	13.2	6.9	73.7
5	12.0	80.8	54.0	13.5	6.3	73.3
6	13.5	76.5	55.7	11.1	4.7	71.5

Reaction conditions: LA 1.5 mmol, Zn 10 mmol, Co 4 mmol, 250 °C, 2 h. Reprinted from Royal Society of Chemistry (Ren et al., 2017)

The effects of temperature and reaction time on the conversion of LA into PA were shown in Fig. 5.12. During the temperature range of 175–250 °C, a big change in the yield of PA was observed. However, the higher reaction temperature was not investigated because the tolerated temperature of the Teflon reactor was 250 °C. But it was possible that a higher PA yield might be attained at more than 250 °C. As a result, the LA was hard to be converted at low temperature 175 °C even after prolonged reaction time, but the effect of reaction time visibly emerged at more than 200 °C, which meant that the reaction only could be motivated at certain temperature. When time was 2 h, the maximum value of 58.8% at 250 °C was achieved. Besides, HPLC chromatogram of liquid samples in Fig. 5.13 showed that yield of PA further increased which yield of LA was further decreased with the time increased.

Finally, mass balance was conducted at the optimal conditions by total organic carbon (TOC). The result showed that approximately 98% organic carbon was converted into liquid products. The gaseous samples were detected by TCD and only H_2 and trace amount of CH_4 and CO_2 were found.

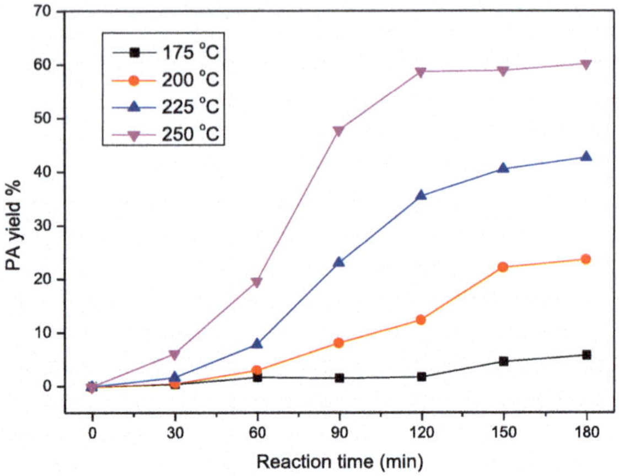

Fig. 5.12 Effects of temperature and time (LA 1.5 mmol, Zn 10 mmol, Co 4 mmol, H_2O 7.5 mL). Reprinted from Royal Society of Chemistry (Ren et al., 2017)

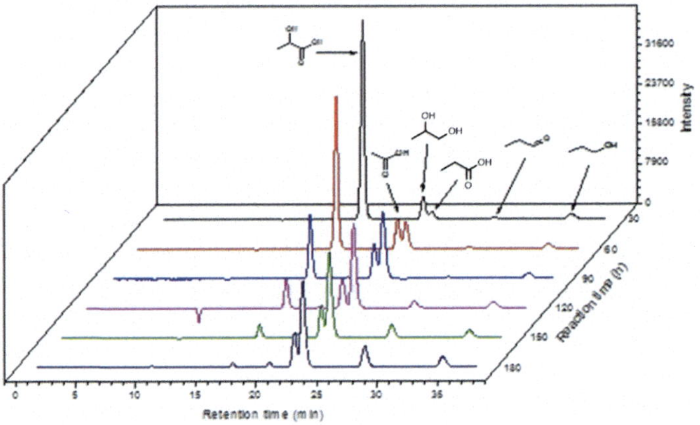

Fig. 5.13 HPLC chromatogram of liquid samples at different time (LA 1.5 mmol, Zn 10 mmol, Co 4 mmol, H_2O 7.5 mL, 250 °C). Reprinted from Royal Society of Chemistry (Ren et al., 2017)

5.3.4 *Recyclability of Co Catalyst*

Recyclability of catalyst Co was investigated at the optimal reaction conditions. Figure 5.14 showed a simple and economical method for the separation of Co from the residue was designed according to their densities. Most of formed ZnO can be recycled after simple filtering. As shown in Fig. 5.15 and Table 5.6, the results indicated that Co catalyst still maintained a high catalytic activity after four cycles for the PA production. XRD patterns of recycled Co catalyst in Fig. 5.16 and SEM images of Co catalyst in Fig. 5.17 did not found the obvious change before and after the reaction. Though yield of PA decreased a little to 56.0%, it might be due to in situ ZnO adsorbed on the surface of Co catalyst to decrease the surface area during the reaction.

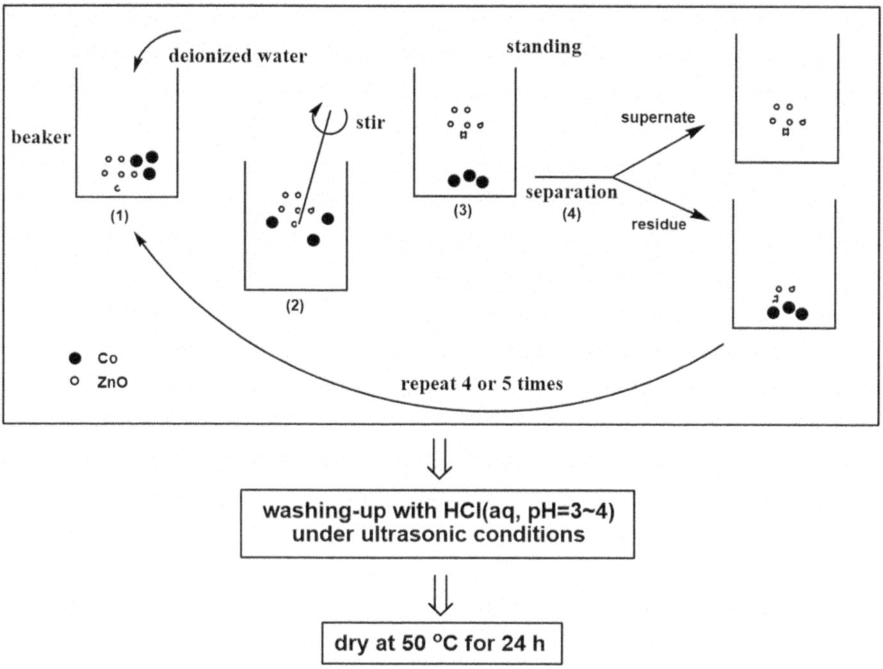

Fig. 5.14 Recovery process of Co catalyst from the solid residues. Reprinted from Royal Society of Chemistry (Ren et al., 2017)

Fig. 5.15 Recyclability of Co catalyst (LA 1.5 mmol, Zn 10 mmol, Co 4 mmol, H_2O 7.5 mL, 250 °C, 2 h). Reprinted from Royal Society of Chemistry (Ren et al., 2017)

5.3.5 *Role of in Situ ZnO*

To investigate the role of in situ ZnO generated by the oxidation of Zn in water, the conversion reaction of LA was performed with gaseous hydrogen as the reductant

Table 5.6 Recyclability of Co catalyst

Entry	Catalyst	Yield of PA (%)
1	Fresh	58.8
2	Reuse 1	60.6
3	Reuse 2	57.9
4	Reuse 3	56.0
5	Reuse 4	58.7

Reaction conditions: LA 1.5 mmol, Zn 10 mmol, Co 4 mmol, H_2O 7.5 mL, 250 °C, 2 h. Reprinted from Royal Society of Chemistry (Ren et al., 2017)

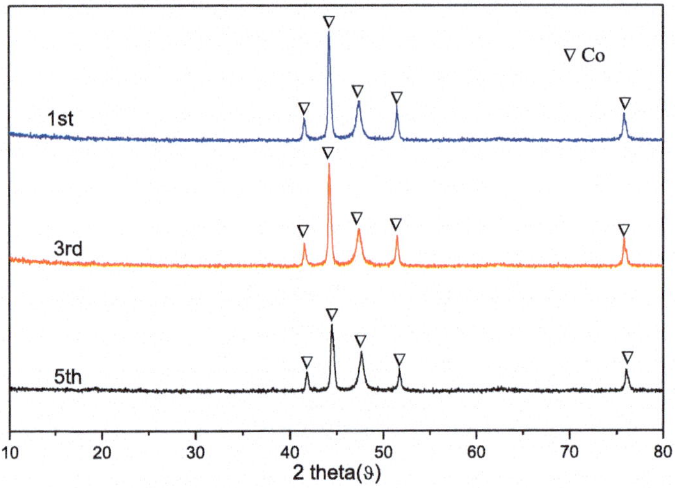

Fig. 5.16 XRD patterns of recycled Co (LA 1.5 mmol, Zn 10 mmol, Co 4 mmol, H_2O 7.5 mL, 250 °C, 2 h). Reprinted from Royal Society of Chemistry (Ren et al., 2017)

Fig. 5.17 SEM images of Co catalyst. Left: fresh, right: after four cycles. Reprinted from Royal Society of Chemistry (Ren et al., 2017)

Table 5.7 Study on the role of in situ ZnO

Entry	Co (mmol)	ZnO	Yield of PA (%)
1	Without	With	0
2	With	Without	30.2
3	With	With	62.5

Reaction conditions: LA 1.5 mmol, ZnO 10 mmol, Co 4 mmol, H_2 1.10 MPa, 250 °C, 2 h. Reprinted from Royal Society of Chemistry (Ren et al., 2017)

instead of in situ hydrogen in the presence/absence of ZnO, the results was summarized in Table 5.7. No desired PA was observed when only ZnO was used, indicating that ZnO did not exhibit catalytic performance (entry 1). The absence of ZnO only gave the PA in 30.2% yield, while yield of PA was remarkably increased to 62.5% with the addition of ZnO (entries 2–3). Therefore, it is suggested that in situ ZnO can enhance the catalytic activity of Co catalyst and their synergetic effect is effective for the transformation of LA into PA in water.

5.3.6 Scope of Biomass Carbohydrates for PA Production

Further expanding the scope of the feedstock was conducted with various biomass carbohydrates. With the conditions reported previously in the first step process (Huo et al., 2014) and the optimum reaction conditions in the second step process from above, we conducted the two-step reactions with glucose, cellulose and starch to produce PA, and the results were summarized in Table 5.8. Maximum PA yields of 16.5% from glucose, 5.2% from cellulose and 15.4% from starch were obtained using the two-step process.

We next carried out the conversion from glycolic acid (GA) to acetic acid (AA) could happen by this method. The reaction of GA was conducted in the presence of 10 mmol Zn and 4 mmol Co in water at 250 °C for 2 h and gave the expected AA and ethylene glycol (EG) in 17.2% and 9.5% yields, respectively (Fig. 5.18).

5.4 Conclusions

We developed a non-fermentative process for PA production from LA and biomass over Co catalyst in hydrothermal conditions. Co showed significantly catalytic activity and gave the yield of 58.8% in the presence of 4 mmol Co and 10 mmol Zn in H_2O at 250 °C for 2 h. The result indicated that Co and ZnO had a synergetic effects on the production of PA from LA. The Recyclability of Co catalyst proved that Co could be repeatedly used four times without the loss of its activity. Biomass carbohydrates such as glucose, cellulose and starch, could also produce the product PA with

Table 5.8 Co-catalyzed conversion of biomass carbohydrates into PA[a]

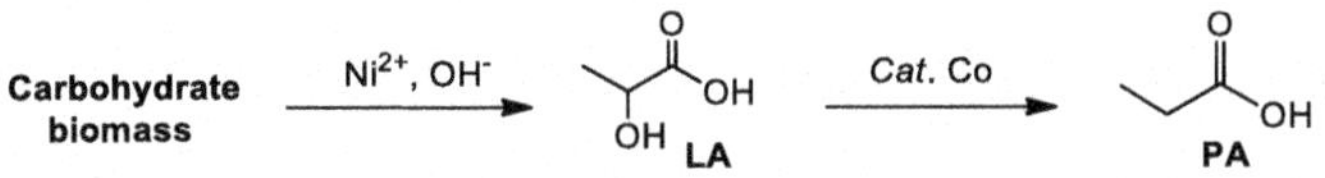

Entry	Carbohydrate	Yield of products				
		PA	AA	PG	NPA	Others[b]
1	Glucose	16.5	2.9	8.5	6.3	n.d
2	Cellulose	5.2	4.2	3.4	2.2	n.d
3	Starch	15.4	3.8	10.1	5.2	n.d
4	Fructose	6.4	7.3	15.6	n.d	n.d
5	Saccharose	5.8	7.3	14.4	n.d	n.d

[a] Reaction condition: (1) 0.01 M NaOH, 0.01 M Ni^{2+}, 0.035 g carbohydrates, 2 mL H_2O, 300 °C, 1 min, and then directly (2) 0.76 mmol Co and 1.9 mmol Zn, 300 °C, 60 min. [b] n.d: not determined. Here the SUS316 reactor was used for the reactions. Reprinted from Royal Society of Chemistry (Ren et al., 2017)

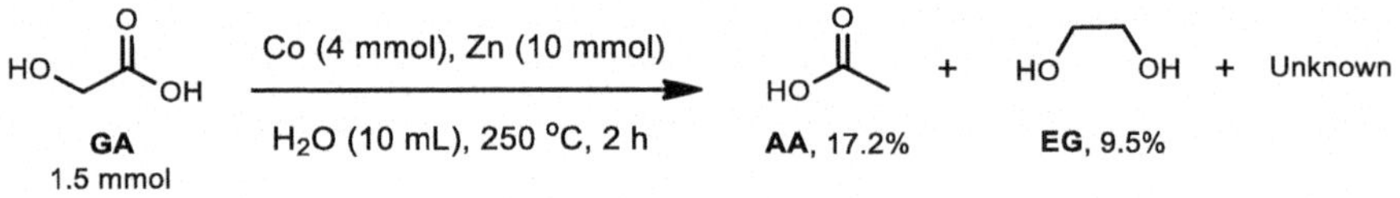

Fig. 5.18 Conversion of glycolic acid (GA). Reprinted from Royal Society of Chemistry (Ren et al., 2017)

two-step process. This method can be also used for the conversion from glycolic acid (GA) to acetic acid (AA) and ethylene glycol. The present study afforded a potential and economical utilization of biomass carbohydrates into PA by the novel chemical method.

References

Bond, J. Q., Upadhye, A. A., Olcay, H., Tompsett, G. A., Jae, J., Xing, R., Alonso, D. M., Wang, D., Zhang, T., & Kumar, R. (2014). Production of renewable jet fuel range alkanes and commodity chemicals from integrated catalytic processing of biomass. *Energy & Environmental Science, 7*, 1500–1523.

Boyaval, P., & Corre, C. (1995). Production of propionic acid. *Le Lait, 75*, 453–461.

Chen, X., Chew, S. L., Kerton, F. M., & Yan, N. (2014). Direct conversion of chitin into a N-containing furan derivative. *Green Chemistry, 16*, 2204–2212.

Corma, A., Iborra, S., & Velty, A. (2007). Chemical routes for the transformation of biomass into chemicals. *Chemical Reviews, 107*, 2411–2502.

Dauenhauer, P. J., & Huber, G. W. (2014). Biomass at the shale gas crossroads. *Green Chemistry, 16*, 382–383.

Dietz, D., Sabra, W., & Zeng, A.-P. (2013). Co-cultivation of Lactobacillus zeae and Veillonella criceti for the production of propionic acid. *AMB Express, 3*, 1–9.

Ding, D., Wang, J., Xi, J., Liu, X., Lu, G., & Wang, Y. (2014). High-yield production of levulinic acid from cellulose and its upgrading to γ-valerolactone. *Green Chemistry, 16*, 3846–3853.

Dishisha, T., Ibrahim, M. H., Cavero, V. H., Alvarez, M. T., & Hatti-Kaul, R. (2015). Improved propionic acid production from glycerol: Combining cyclic batch-and sequential batch fermentations with optimal nutrient composition. *Bioresource Technology, 176*, 80–87.

Feng, X., Chen, F., Xu, H., Wu, B., Li, H., Li, S., & Ouyang, P. (2011). Green and economical production of propionic acid by Propionibacterium freudenreichii CCTCC M207015 in plant fibrous-bed bioreactor. *Bioresource Technology, 102*, 6141–6146.

Gallezot, P. (2012). Conversion of biomass to selected chemical products. *Chemical Society Reviews, 41*, 1538–1558.

Gao, P., Li, G., Yang, F., Lv, X.-N., Fan, H., Meng, L., & Yu, X.-Q. (2013). Preparation of lactic acid, formic acid and acetic acid from cotton cellulose by the alkaline pre-treatment and hydrothermal degradation. *Industrial Crops and Products, 48*, 61–67.

Gao, X., Chen, X., Zhang, J., Guo, W., Jin, F., & Yan, N. (2016). Transformation of chitin and waste shrimp shells into acetic acid and pyrrole. *ACS Sustainable Chemistry & Engineering, 4*, 3912–3920.

Hsu, S. T., & Yang, S. T. (1991). Propionic acid fermentation of lactose by Propionibacterium acidipropionici: Effects of pH. *Biotechnology and Bioengineering, 38*, 571–578.

Huber, G. W., Iborra, S., & Corma, A. (2006). Synthesis of transportation fuels from biomass: Chemistry, catalysts, and engineering. *Chemical Reviews, 106*, 4044–4098.

Huo, Z., Fang, Y., Ren, D., Zhang, S., Yao, G., Zeng, X., & Jin, F. (2014). Selective conversion of glucose into lactic acid with transition metal ions in diluted aqueous NaOH solution. *ACS Sustainable Chemistry & Engineering, 2*, 2765–2771.

Jiang, L., Cui, H., Zhu, L., Hu, Y., Xu, X., Li, S., & Huang, H. (2015). Enhanced propionic acid production from whey lactose with immobilized Propionibacterium acidipropionici and the role of trehalose synthesis in acid tolerance. *Green Chemistry, 17*, 250–259.

Jin, F., & Enomoto, H. (2009). Hydrothermal conversion of biomass into value-added products: Technology that mimics nature. *BioResources, 4*, 704–713.

Jin, F., & Enomoto, H. (2011). Rapid and highly selective conversion of biomass into value-added products in hydrothermal conditions: Chemistry of acid/base-catalysed and oxidation reactions. *Energy & Environmental Science, 4*, 382–397.

Lewis, V. P., & Yang, S.-T. (1992). Propionic acid fermentation by Propionibacterium acidipropionici: Effect of growth substrate. *Applied Microbiology and Biotechnology, 37*, 437–442.

Liu, Y., Zhang, Y.-G., Zhang, R.-B., Zhang, F., & Zhu, J. (2011). Glycerol/glucose co-fermentation: One more proficient process to produce propionic acid by Propionibacterium acidipropionici. *Current Microbiology, 62*, 152–158.

Martínez-Campos, R., & de la Torre, M. (2002). Production of propionate by fed-batch fermentation of Propionibacterium acidipropionici using mixed feed of lactate and glucose. *Biotechnology Letters, 24*, 427–431.

Miao, G., Zhu, C., Wang, J., Tan, Z., Wang, L., Liu, J., Kong, L., & Sun, Y. (2015). Efficient one-pot production of 1, 2-propanediol and ethylene glycol from microalgae (Chlorococcum sp.) in water. *Green Chemistry, 17*, 2538–2544.

Ren, D., Fu, J., Li, L., Liu, Y., Jin, F., & Huo, Z. (2016a). Efficient conversion of biomass-derived furfuryl alcohol to levulinate esters over commercial α-Fe_2O_3. *RSC Advances, 6*, 22174–22178.

Ren, D., Song, Z., Li, L., Liu, Y., Jin, F., & Huo, Z. (2016b). Production of 2, 5-hexanedione and 3-methyl-2-cyclopenten-1-one from 5-hydroxymethylfurfural. *Green Chemistry, 18*, 3075–3081.

Ren, D., Huo, Z., Xiao, J., Jin, F., Wang, T., & Yao, G. (2017). Chemoselective synthesis of propionic acid from biomass and lactic acid over a cobalt catalyst in aqueous media. *Green Chemistry, 19*, 1308–1314.

Schmidt, L. D., & Dauenhauer, P. J. (2007). Hybrid routes to biofuels. *Nature, 447*, 914–915.

Testova, L., Nieminen, K., Penttilä, P. A., Serimaa, R., Potthast, A. & Sixta, H. (2014). Cellulose degradation in alkaline media upon acidic pretreatment and stabilisation. *Carbohydrate Polymers*, 100, 185–194.

Wang, F.-F., Liu, C.-L., & Dong, W.-S. (2013). Highly efficient production of lactic acid from cellulose using lanthanide triflate catalysts. *Green Chemistry, 15*, 2091–2095.

Wang, J., Yao, G., Wang, Y., Zhang, H., Huo, Z., & Jin, F. (2015). A novel Pd/C-catalyzed conversion of glucose to 1, 2-propanediol by water splitting with Zn. *RSC Advances, 5*, 51435–51439.

Xiao, J., Huo, Z., Ren, D., Zhang, S., Luo, J., Yao, G., & Jin, F. (2015). A novel approach for 1, 2-propylene glycol production from biomass-derived lactic acid. *Process Biochemistry, 50*, 793–798.

Xu, L., Huo, Z., Fu, J., & Jin, F. (2014). Highly efficient conversion of biomass-derived glycolide to ethylene glycol over CuO in water. *Chemical Communications, 50*, 6009–6012.

Chapter 6
A Novel Approach for 1,2-Propylene Glycol Production from Biomass-Derived Lactic Acid

6.1 Introduction

The development and utilization of renewable and clean biomass resource have attracted much attention because of the energy shortages and environmental pollution, caused by the over-consumption of non-renewable fossil fuel. As an abundant, renewable, low-cost, and less pollution resource, biomass resource has its own advantages in comparison with fossil resource. Previous researches showed that the application of biomass as the raw material for the production of chemical stocks has a promising future to replace the petroleum-derived materials (Bond et al., 2010; Chheda et al., 2007; Hermann et al., 2007; Huber & Corma, 2007).

1,2-propylene glycol, an important multi-purpose chemical that has an extensive use as anti-freeze agent, chemical application, consumer products and foods (Cameron et al., 1998; Saxena et al., 2010), is industrially produced by the hydration of propylene oxide that derived from petroleum-derived propylene or by the hydroperoxide process (Szmant, 1989). Due to its industrial application value, the synthesis of 1,2-propylene glycol is attracting interest in recent years, 1,2-propylene glycol and other polyols can be effectively obtained from cellulose and chitin derived sugars (Bobbink et al., 2015; Yan & Dyson, 2013). Lactic acid as an important biomass-derived compound, which has high reactivity because of the connection of hydroxyl and carboxy groups, can be obtained by the fermentation of agricultural crops and biomass (Litchfield, 1996; Lipinsky & Sinclair, 1986; Wee et al., 2006), the catalytic hydrogenation of lactic acid into 1,2-propylene glycol is becoming a feasible and commercial method to use the renewable biomass energy (Eq. 6.1). For example, in the 1930s, Adkins et al. reported the conversion of neat ethyl lactate into 1,2-propylene glycol over Raney Ni in exceeding 80% yield at very high hydrogen pressures (>25 MPa) (Adkins & Billica, 1948; Bowden & Adkins, 1934). Broadbent et al. performed the catalytic hydrogenation of free lactic acid into 1,2-propylene glycol in 84% yield over unsupported rhenium black at 27 MPa hydrogen pressure (Broadbent et al., 1959). Dumesic et al. converted lactic acid into 1,2-propylene glycol in 88%

Z. Huo, *Diverse Hydrogen Sources for Biomass-derivatives Conversion*,
https://doi.org/10.1007/978-981-99-1673-3_6

yield over a silica-supported copper catalyst with a lower 0.10 MPa hydrogen pressure (Cortright et al., 2002). Miller et al. reported the catalytic hydrogenation of lactic acid into 1,2-propylene glycol in 90% yield over ruthenium supported on activated carbon at 14 MPa hydrogen pressure (Zhang et al., 2001). More recently, Baeck et al. reported a novel method for the synthesis of 1,2-propylene glycol from lactic acid over ruthenium catalyst at hydrogen pressure 50 bar and gave middle yields (Jang et al., 2011, 2013). However, these methods still face academic and industrial challenges, such as use of high pressure gases hydrogen, expensive synthesized catalysts and longer reaction time, etc. Thus, the development of an efficient process for the production of 1,2-propylene glycol from lactic acid is highly desired.

Recently, hydrothermal treatment in the conversion of biomass into high value-added chemicals has caught much attention and gradually becomes a novel way of utilizing the renewable energy because of its unique merits (dielectric constant, density and caloric capacity, etc.) (Akiya & Savage, 2002; Duan & Savage, 2011; Fang et al., 2008; He et al., 2010; Jin & Enomoto, 2009; Tian et al., 2007; Wang et al., 2013). Previously, our researches found that CO_2 could be reduced to formic acid and/or methanol by in situ formed hydrogen by the oxidation of Zn in water (Huo et al., 2012; Jin et al., 2011). More recently, we reported that a highly efficient synthesis of ethylene glycol from biomass-derived glycolide over CuO in high yield with high selectivity (Xu et al., 2014). Inspired by the findings above, this led us to an idea that 1,2-propylene glycol could be synthesized catalytically from lactic acid in high temperature water (Eq. 6.2). The purpose of this study was to present an efficient method for the production of 1,2-propylene glycol from lactic acid with high selectivity and productivity (Fig. 6.1).

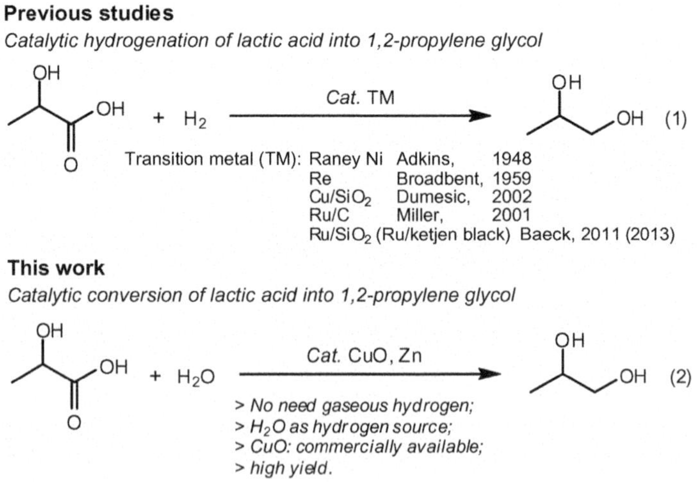

Fig. 6.1 Previous studies and our concept of lactic acid conversion. Reprinted from Elsevier (Xiao et al., 2015)

6.2 Material and Methods

6.2.1 General Information

Lactic acid (DL, ~90%T) was obtained from Sigma-Aldrich Chemical Reagent Co., Ltd. Zn powder (AR) was obtained from Aladdin Chemical Reagent Co. Ltd. CuO powder (200-mesh, 99.7%), other metals and metal oxides were obtained from Sinopharm Chemical Reagent Co. Ltd, China. The standard reagents 1,2-propylene glycol (J&K Chemical, > 99.8%), Hydroxyacetone (Alfa Aesar, > 95%), Acetic acid (Sigma-Aldrich, > 99.7%), Propionic acid (Sigma- Aldrich, > 99.7%) and lactic acid (Alfa Aesar, 1.0 N Standardized lactic acid solution) were used in the qualitative analysis of liquid example.

6.2.2 Product Analysis

The qualitative analysis of obtained liquid samples was based on the results of GC–MS (Agilent 7890 system, equipped with 5985C inert MSD with triple-axis detector) while the quantitive analysis of 1,2-propylene glycol and byproducts was conducted by HPLC (Agilent 1200 system, equipped with VW and RI detectors). The solution was also tested by ICP to measured the ion concentration. Solid sample was dried in an oven at 60 °C for 24 h after filtered cleaning with distilled water. Then precipitate was analyzed by XRD (Bruker D8 Advance X-Ray Diffractometer, equipped with Cu Kα radiation: $\lambda = 1.5406$ Å, scanning rate: 0.02° s^{-1}, 2θ ranges: 10–80°).

The lactic acid conversion ($LA_{Conv.}$), product i yield (Y_i) and selectivity to product i (S_i) were calculated using following equations. The yield was obtained from three replicate experiments, and the relative error was less than 5%.

$$LA_{Conv.}(\%) = \left(1 - \frac{mole of lactic acid at particular time}{initial mole of lactic acid}\right) \times 100 \quad (6.1)$$

$$Y_i\,(\%) = \left(\frac{mole\ of\ product\ i}{initial\ mole\ of\ lactic\ acid}\right) \times 100 \quad (6.2)$$

$$S_i\,(\%) = \left(\frac{mole\ of\ product\ i}{mole\ of\ all\ products\ formed}\right) \times 100 \quad (6.3)$$

6.2.3 Experimental Procedure for the Synthesis of 1,2-Propylene Glycol from Lactic Acid

All reactions were performed in a batch Teflon reactor whose inner volume is 30 mL. The procedure was as follows. Firstly, the desired concentration of lactic acid solution was prepared according to water filling. Then, the desired catalysts and reductants were loaded into the reactor. The reactor was put into the preheated oven for 20 min for temperature rising to the desired temperature. Afterward, the reactor was moved from the oven and cooled with a electric fan. The reaction time was defined as the time when the reactor's temperature was up to 250 °C. The water filling was defined as the ratio of volume of the lactic acid solution put into the reactor and inner volume of reactor. Finally, after cooling off, liquid sample was collected and filtered with 0.45 μm Syringe Filter. Solid sample was collected and washed with deionized water and ethanol several times to remove impurities and dried in the oven at 50 °C for 24 h.

6.3 Results and Discussion

The reactions of lactic acid to 1,2-propylene glycol were carried out to investigate the optimum conditions. Some parameters, such as the activity of catalysts, catalyst loading, reductant, reductant loading, water filling, reaction temperature, reaction time, effect of pure hydrogen, were investigated systematically in details. The results obtained from the investigation are discussed as follows.

6.3.1 Catalyst Screening

Initially, to investigate the effect of different metals or metal oxides on the conversion of lactic acid, we conducted the experiment in the presence of Cu and Zn with 25% water filling at 250 °C for 2 h. The results indicated that 1,2-propylene glycol was the main product with trace amounts of acrylic acid, acetic acid and propionic acid as shown in Fig. 6.2. As illustrated in Table 6.1, we can see that no desired 1,2-propylene glycol was obtained in the absence of catalyst and/or reductant in the transformation (entries 1–3). Among all catalysts we tested, CuO exhibited excellent catalytic activity for the conversion of lactic acid to 1,2-propylene glycol with conversion of 88% and yield of 86% (entry 8). The use of Cu, Cu_2O, Ni, Ni_2O_3, Fe and TiO_2 gave a low conversion rate and low yields (entries 4–6, 9–11). Other catalysts, such as Mo, Fe_2O_3, ZrO_2 and SiO_2, were ineffective and no desired 1,2-propylene glycols were achieved (entries 7, 12–14). The present study indicated that CuO had a significant influence on the formation of 1,2-propylene glycol from lactic acid.

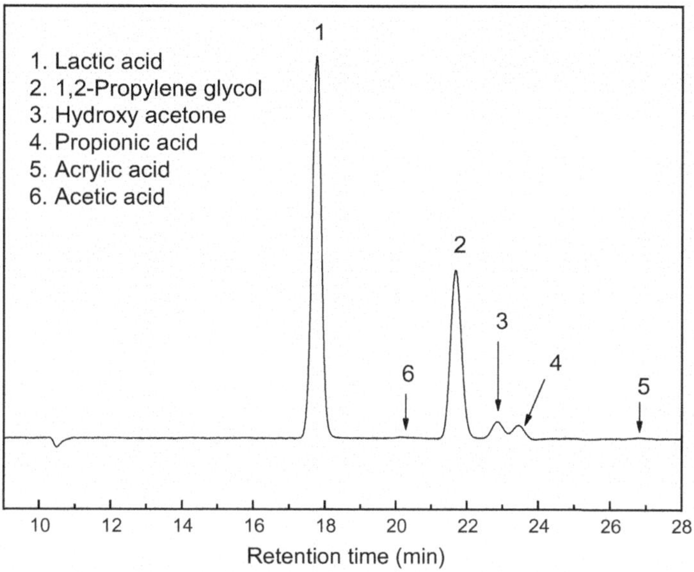

Fig. 6.2 Chromatogram obtained by HPLC of the liquid sample after the reaction (temp: 250 °C, water filling: 25%, time: 2 h, Cu: 5 mmol, Zn: 25 mmol). Reprinted from Elsevier (Xiao et al., 2015)

As the XRD pattern of solid residues shown in Fig. 6.3a, we found that Cu still existed while Zn was oxidized into ZnO, it was obviously that Cu acted as a catalyst and Zn acted as a reductant in this process. According to the result presented in Fig. 6.3b, the Cu instead of CuO was observed in the solid sample. We thought that CuO shows an excellent catalytic activity during the conversion reaction before being reduced into Cu by in situ formed hydrogen. Our previous work (Huo et al., 2012) had confirmed that CuO could be reduced completely into Cu by in situ formed hydrogen by the oxidation of Zn in water or 1,2-propylene glycol. As shown in Fig. 6.4, we investigated the oxidation of Zn, and found that Zn was oxidized quickly to ZnO with the increase of reaction time from 1 to 40 min until the complete oxidation at 40 min. Furthermore, little Cu was detected in the solid sample, which indicated CuO was just reduced into Cu after 40 min of reaction. Thus, CuO and Cu maintaining coexistence from the beginning of the present stage. But the amount of CuO decreased gradually, whereas the amount of Cu increased with the increase of reaction time until the complete reduction into Cu. However, in all cases, we cannot detect the peaks of CuO from XRD, this may be because of the peak intensity of Zn, Cu and ZnO is stronger than that of CuO peaks (In addition, we measured Cu ion concentration in liquid sample by ICP after the reaction, and only a small amount of Cu ion (less than 1 ppm) was found in liquid sample. This suggested that most of Cu was presented in the solid residue.

Next, optimizing the amount of CuO from 0 to 15 mmol to obtain a high yield of 1,2-propylene glycol. As shown in Fig. 6.5, the 1,2-propylene glycol yield increased

Table 6.1 Effect of catalysts on the conversion of lactic acid[a]

Entry	Catalyst	Reductant	Lactic acid conversion (%)	Yield (%)[b,c]			
				1,2-PG	HA	AA	PA
1	–	–	0	0	0	0	0
2	CuO	–	21	0	0	2	< 1
3	–	Zn	0	0	0	0	0
4	Cu	Zn	46	40	4	< 1	2
5	Ni	Zn	26	21	4	< 1	1
6	Fe	Zn	7	7	< 1	< 1	< 1
7	Mo	Zn	1	0	0	< 1	1
8	CuO	Zn	88	86	< 1	< 1	2
9	Cu_2O	Zn	45	36	3	< 1	3
10	Ni_2O_3	Zn	24	10	0	8	6
11	TiO_2	Zn	6	< 1	0	< 1	6
12	Fe_2O_3	Zn	4	0	4	< 1	< 1
13	ZrO_2	Zn	1	0	< 1	1	0
14	SiO_2	Zn	1	0	1	< 1	0

Reprinted from Elsevier (Xiao et al., 2015)
[a] Reaction condition: lactic acid: 0.83 mmol; catalyst: 5 mmol; reductant: 25 mmol;temp: 250°C; time: 2 h; water filing: 25%. [b] Abbreviations: PA: Propionic acid, AA: Acetic acid, HA: Hydroxyacetone, PG: 1,2-propylene glycol, LA: Lactic acid. [c] The yield is calculated as the percentage of 1,2-propylene glycol to the initial lactic acid on the carbon base

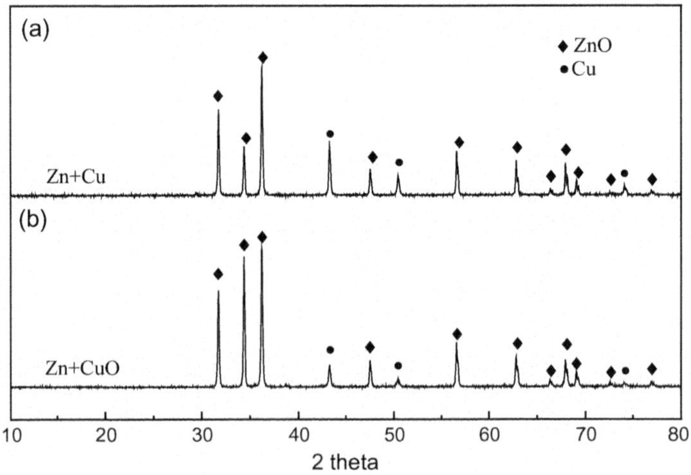

Fig. 6.3 XRD pattern of soil residual after the reaction with Cu and CuO (lactic acid: 0.83 mmol; catalyst: 5 mmol; Zn: 25 mmol; temp: 250 °C; time: 2 h; water filling: 25%). Reprinted from Elsevier (Xiao et al., 2015)

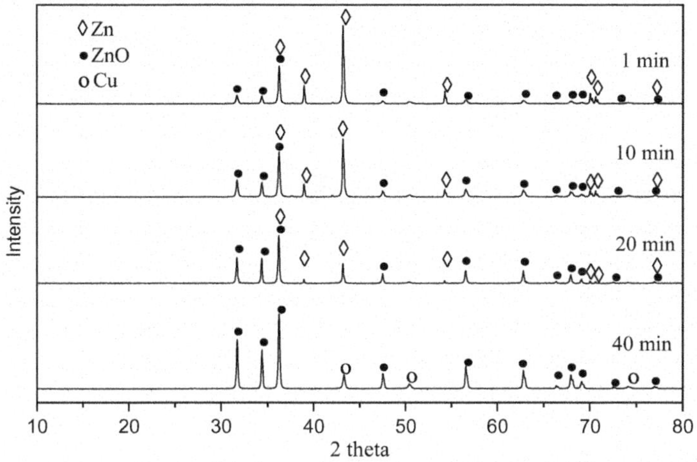

Fig. 6.4 XRD patterns and the conversion of Zn into ZnO (lactic acid: 0.83 mmol; CuO: 5 mmol; Zn: 25 mmol; water filling: 25%; temp: 250 °C). Reprinted from Elsevier (Xiao et al., 2015)

significantly with an increase in the amount of CuO, and reached the maximum value of 86% when CuO was 5 mmol. Then the yields decreased slowly from 5 and 12 mmol, and a rapidly decreasing was observed when CuO amount increased up to 14 mmol. This is probably the result of the decomposition of 1,2-propylene glycol. A relative experiment was conducted to validate this assumption, we found that concentration of 100 mmol/L 1,2-propylene glycol was decomposed and decreased to 90 mmol/L in the presence of 5 mmol Cu and 25 mmol ZnO with 25% water filling at 250 °C for 2 h. Therefore, 5 mmol CuO is suitable for the formation of 1,2-propylene glycol from lactic acid in the process.

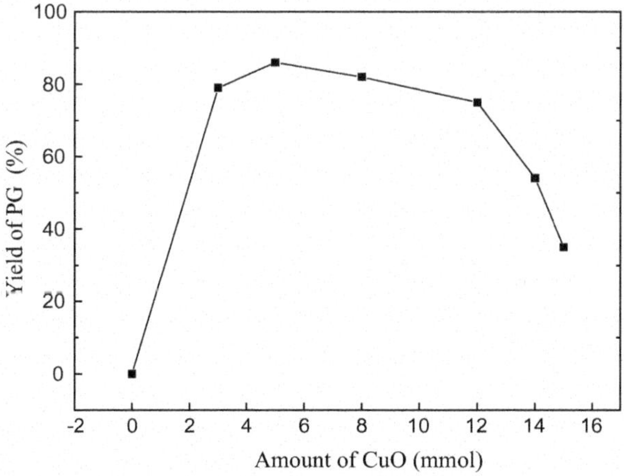

Fig. 6.5 Effect of the amount of CuO on the production of 1,2-propylene glycol (lactic acid: 0.83 mmol (75 mg); Zn: 25 mmol; temp: 250 °C; time: 2 h; water filling: 25%). Reprinted from Elsevier (Xiao et al., 2015)

6.3.2 Optimization of Reductants

Several active metals, such as Al, Mg, Fe and Zn, were investigated to search for a suitable reductant for the formation of 1,2-propylene glycol. As shown in Table 6.2 and Fig. 6.6, it can be seen that Al, Mg and Fe were ineffective, and had no reaction or gave lower yields (entries 1–3), while Zn powder had the much higher reactivity for the conversion of lactic acid to 1,2-propylene glycol compared to other metals, and the yield of 86% for 1,2-propylene glycol was obtained (entry 4). Interestingly, the yield of 27% for acetic acid was obtained with Mg as a reductant (entry 2). According to the results of XRD (see Figs. 6.7 and 6.8), we suggested that oxidations of Al, Mg and Fe were not complete, which led to a small amount of hydrogen production by the oxidation of these metals during the reaction. That is why the reactions did not proceed or provided a lower yield of 1,2-propylene glycol when these metals were used as reductants.

Table 6.2 Effect of reductants on the conversion of lactic acid

Entry	Catalyst	Lactic acid conversion (%)	Yield (%)			
			1,2-PG	HA	AA	PA
1	Al	63	0	1	3	6
2	Mg	57	2	0	27	1
3	Fe	43	23	0	1	7
4	Zn	88	86	2	< 1	< 1

Reprinted from Elsevier (Xiao et al., 2015)
Reaction condition: lactic acid: 0.83 mmol; CuO: 5 mmol; reductant: 25 mmol; temp: 250 °C; time: 2 h; water filing: 25%

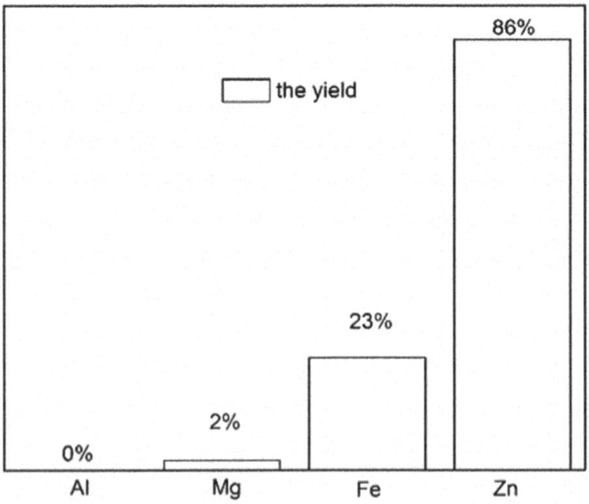

Fig. 6.6 Effect of reductants on the production of propylene glycol (lactic acid: 0.83 mmol (75 mg); CuO: 5 mmol; reductant: 25 mmol; temp: 250 °C; time: 2 h; water filling: 25%). Reprinted from Elsevier (Xiao et al., 2015)

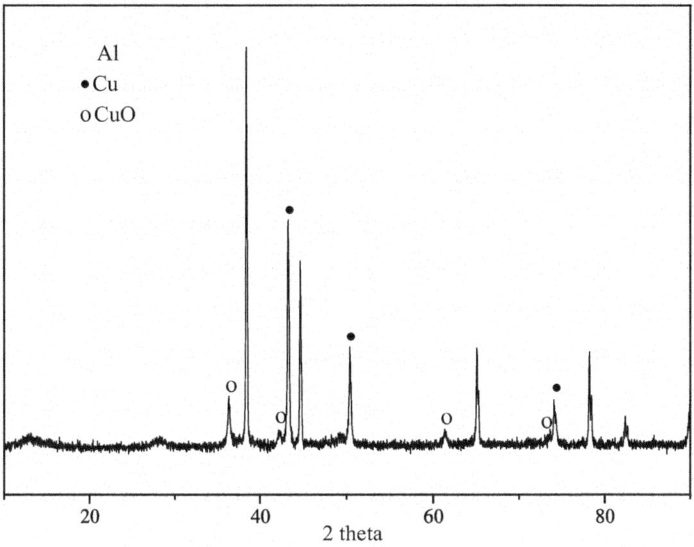

Fig. 6.7 The XRD pattern of precipitates with Fe as reductant (temp: 250 °C; time: 2 h; CuO: 5 mmol; Al: 25 mmol; water filling: 25%). Reprinted from Elsevier (Xiao et al., 2015)

Generally, increasing hydrogen production by the oxidation of Zn contributed to the formation of 1,2-propylene glycol from lactic acid. Experiments were performed to investigate the effect of the amount of Zn by changing from 0 to 40 mmol. As illustrated in Fig. 6.9, the yield of 1,2-propylene glycol almost remained unchangeable from 0 to 5 mmol, and then increased significantly until Zn amount increased to 30 mmol. The maximum value of 90% was obtained when Zn amount was 30 mmol. The formation of 1,2-propylene glycol reached to a plateau when Zn exceeded 30 mmol, which indicates that a higher hydrogen partial pressure did not mean higher yield. Therefore, 30 mmol Zn was chosen for the following experiments.

6.3.3 *Effect of Various Parameters*

6.3.3.1 Effect of Water Filling

As shown in Fig. 6.10, water filling was changed from 10 to 40% to investigate its effect on the yield of 1,2-propylene glycol. Between the water filling of 10% and 30%, 1,2-propylene glycol yield was almost interestingly constant. The yield of 1,2-propylene glycol was improved a little when water filling reached to 35%, and a maximum value of 93% was achieved. When water filling exceeded 35%, the yield was almost unchangeable. However, experiments at more than 40% water filling were not carried on due to the security factor.

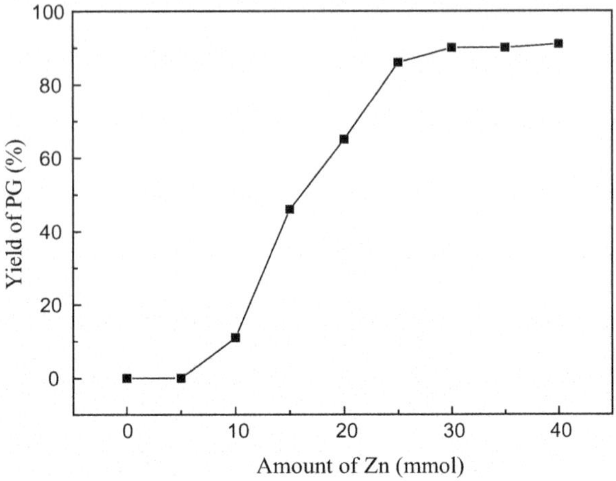

Fig. 6.8 The XRD pattern of precipitates with Mg as reductant (temp: 250 °C; time: 2 h; CuO: 5 mmol; Mg: 25 mmol; water filling: 25%). Reprinted from Elsevier (Xiao et al., 2015)

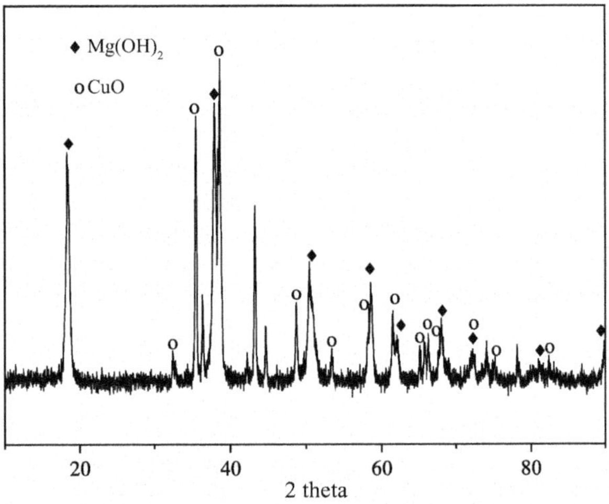

Fig. 6.9 Effect of the amount of Zn on the 1,2-propylene glycol (lactic acid: 0.83 mmol (75 mg); CuO: 5 mmol; temp: 250 °C; time: 2 h; water filling: 25%). Reprinted from Elsevier (Xiao et al., 2015)

6.3.3.2 Effect of Temperature and Time

Figure 6.11a showed the effect of reaction temperature on the conversion of lactic acid to 1,2-propylene glycol. As a result, the yield of 1,2-propylene glycol from lactic acid increased continuously with the temperature changed from 100 to 250 °C. Between 100 and 150 °C, the most obvious increase of the yields of 1,2-propylene glycol could be observed, and the lactic acid conversion to 1,2-propylene glycol slowed down visibly with the temperature increasing from 225 to 250 °C. The maximum value of 93% was achieved at 250 °C. However, 250 °C was the tolerated temperature of the Teflon reactor, we did not continue the optimization at above 250 °C. Hence, the

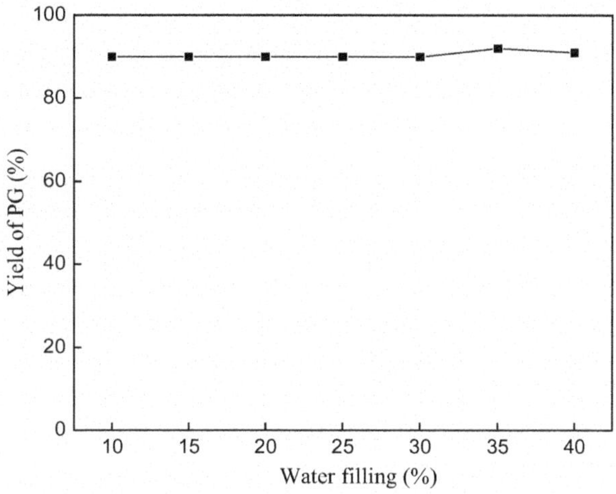

Fig. 6.10 Effect of water filling on the production of 1,2-propylene glycol (lactic acid: 0.83 mmol (75 mg); CuO: 5 mmol; Zn: 30 mmol; temp: 250 °C; time: 2 h). Reprinted from Elsevier (Xiao et al., 2015)

results above suggested that reaction temperature had a significant influence on the production of 1,2-propylene glycol from lactic acid under hydrothermal condition.

Experiments were performed at various reaction time over CuO to investigate the effect of reaction time. As illustrated in Fig. 6.11b, the yield of 1,2-propylene glycol increased significantly with the increase of reaction time in the first 2 h, and led to the highest yield of 93% at 2 h. After 2 h, the yield of 1,2-propylene glycol showed a downward trend. The decreasing yield of 1,2-propylene glycol after 2 h can be attributed to the decomposition of produced 1,2-propylene glycol. Perhaps, the solid residue of Cu and ZnO obtained during the reaction also had some catalytic effect on the decomposition of 1,2-propylene glycol. We confirmed this assumption by the reaction of 1,2-propylene glycol in the presence of 5 mmol Cu and 30 mmol ZnO with 35% water filling at 250 °C for 2 h. The result showed that the concentration

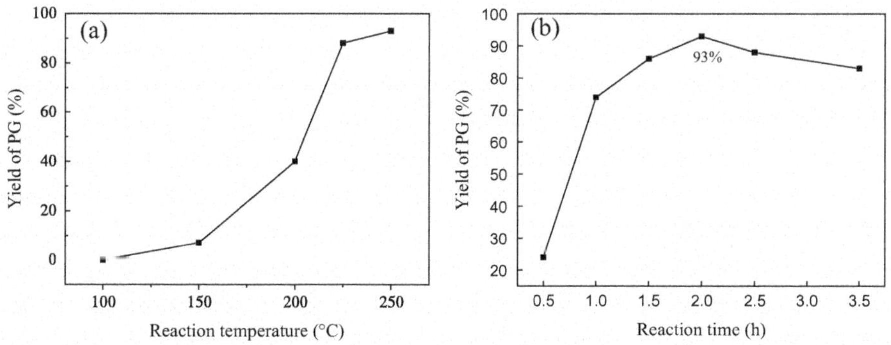

Fig. 6.11 Effect of reaction temperature **a** and time **b** on the production of 1,2-propylene glycol (lactic acid: 0.83 mmol; CuO: 5 mmol; Zn: 30 mmol; water filling: 35%). Reprinted from Elsevier (Xiao et al., 2015)

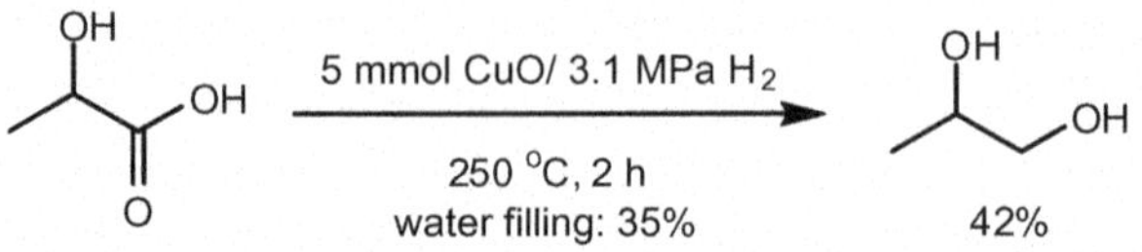

Fig. 6.12 The effect of gases hydrogen on the formation of 1,2-propylene glycol from lactic acid (5 mmol CuO, 3.1 MPa H_2, 35% water filling, 250 °C, 2 h). Reprinted from Elsevier (Xiao et al., 2015)

of 1,2-propylene glycol 100 mmol/L was decreased to 90 mmol/L by HPLC/RI, XRD analysis also indicated that Cu and ZnO still existed in solid residual. The result above suggested that long reaction time has an influence on the yields of 1,2-propylene glycol in the presence of Cu and ZnO obtained in the reaction (Xu et al., 2014).

Finally, a carbon balance was carried out at the optimal condition by TOC and found that approximately 99% lactic acid was converted into liquid products. And less than 1% lactic acid was converted into gaseous products. According to the results of TCD, we found that only hydrogen and trace amounts of other gaseous products were detected.

6.3.4 *Investigation of Gases Hydrogen*

As the results reported in literature (Adkins & Billica, 1948; Bowden & Adkins, 1934; Broadbent et al., 1959; Cortright et al., 2002; Jang et al., 2011, 2013; Zhang et al., 2001) the catalytic hydrogenation of lactic acid gave high yield of 1,2-propylene glycol using gases hydrogen. We investigated the effect of gases hydrogen on the formation of 1,2-propylene glycol from lactic acid under hydrothermal conditions. The reaction was performed in the presence of 5 mmol CuO, 3.1 MPa H_2, 35% water filling at 250 °C for 2 h, and the yield of 42% was much lower than that of 93% yield by in situ formed hydrogen by Zn oxidation in water (Fig. 6.12). The result suggested that gases hydrogen was unsuitable for the formation of 1,2-propylene glycol from lactic acid under hydrothermal conditions.

6.4 Mechanism

Based on literatures reported (Davachi et al., 2012) and results of this study, the pathway for the conversion of lactic acid to 1,2-propylene glycol over CuO was proposed in Fig. 6.13. Firstly, hydrogen was formed by the oxidation of Zn in water (Elliott, 2008; Resende & Savage, 2010). The in situ formed hydrogen was adsorbed on the surface of Cu/CuO/ZnO through the formation of hydrogen bonds between hydrogen and Cu/CuO/ZnO. Also lactic acid was adsorbed on the surface

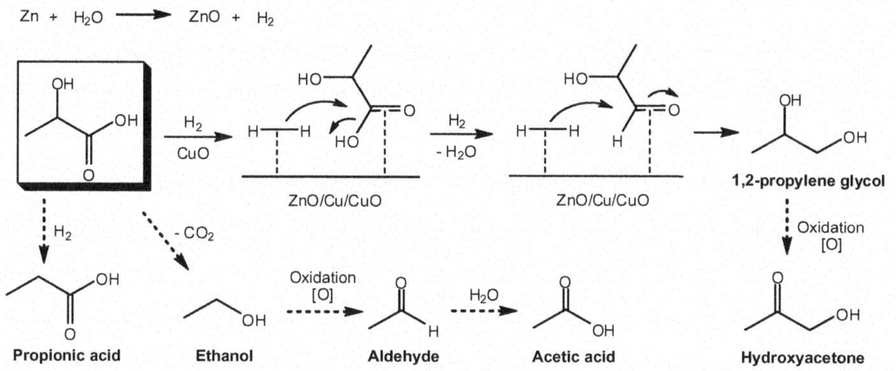

Fig. 6.13 Plausible pathway for the 1,2-propylene glycol production from lactic acid. Reprinted from Elsevier (Xiao et al., 2015)

of Cu/CuO/ZnO by the combiner of Cu/CuO/ZnO with carbonylic C and O atoms (Kieboom & Rantwijk, 1977; Yan et al., 2009). When the C = O bond of lactic acid is further hydrogenated, and desired 1,2-propylene glycol was obtained. The oxidation of 1,2-propylene glycol makes hydroxyacetone in water. Propionic acid is produced via the substitution of lactic acid. Acetic acid is made through decarboxylation and oxidation of the lactic acid.

6.5 Conclusions

We developed an efficient process for 1,2-propylene glycol production from lactic acid over CuO under hydrothermal conditions. CuO showed significantly catalytic activity and gave the yield of 93% in the presence of 5 mmol CuO and 30 mmol Zn with 35% water filling at 250 °C for 2 h. The result indicated that in situ formed hydrogen by the oxidation of Zn in water was efficient for the conversion of lactic acid compared to the use of gases hydrogen. The present study offered a promising method for the efficient conversion of lactic acid into 1,2-propylene glycol under hydrothermal conditions.

References

Adkins, H., & Billica, H. R. (1948). The hydrogenation of esters to alcohols at 25–150°. *Journal of the American Chemical Society, 70*, 3121–3125.

Akiya, N., & Savage, P. E. (2002). Roles of water for chemical reactions in high-temperature water. *Chemical Reviews, 102*, 2725–2750.

Bobbink, F. D., Zhang, J., Pierson, Y., Chen, X., & Yan, N. (2015). Conversion of chitin derived N-acetyl-d-glucosamine (NAG) into polyols over transition metal catalysts and hydrogen in water. *Green Chemistry, 17*, 1024–1031.

Bond, J. Q., Alonso, D. M., Wang, D., West, R. M., & Dumesic, J. A. (2010). Integrated catalytic conversion of γ-valerolactone to liquid alkenes for transportation fuels. *Science, 327*, 1110–1114.

Bowden, E., & Adkins, H. (1934). Hydrogenation of optically active compounds over nickel and copper—Chromium oxide. *Journal of the American Chemical Society, 56*, 689–691.

Broadbent, H. S., Campbell, G. C., Bartley, W. J., & Johnson, J. H. (1959). Rhenium and its compounds as hydrogenation catalysts. III. Rhenium Heptoxide 1, 2, 3. *Journal of Organic Chemistry, 24*, 1847–1854.

Cameron, D., Altaras, N., Hoffman, M., & Shaw, A. (1998). Metabolic engineering of propanediol pathways. *Biotechnology, 14*, 116–125.

Chheda, J. N., Huber, G. W., & Dumesic, J. A. (2007). Liquid-phase catalytic processing of biomass-derived oxygenated hydrocarbons to fuels and chemicals. *Angewandte, 46*, 7164–7183.

Cortright, R., Sanchez-Castillo, M., & Dumesic, J. (2002). Conversion of biomass to 1, 2-propanediol by selective catalytic hydrogenation of lactic acid over silica-supported copper. *Applied Catalysis, b: Environmental, 39*, 353–359.

Davachi, S., Kaffashi, B., & Roushandeh, J. M. (2012). Synthesis and characterization of a novel terpolymer based on L-lactide, glycolide, and trimethylene carbonate for specific medical applications. *Polymers for Advanced Technologies, 23*, 565–573.

Duan, P., & Savage, P. E. (2011). Upgrading of crude algal bio-oil in supercritical water. *Bioresource Technology, 102*, 1899–1906.

Elliott, D. C. (2008). Catalytic hydrothermal gasification of biomass. *Biofuels, Bioproducts and Biorefining, 2*, 254–265.

Fang, Z., Sato, T., Smith Jr., R. L., Inomata, H., Arai, K. & Kozinski, J. A. 2008. Reaction chemistry and phase behavior of lignin in high-temperature and supercritical water. *Bioresource Technology, 99*, 3424–3430.

He, C., Tian, G., Liu, Z., & Feng, S. (2010). A mild hydrothermal route to fix carbon dioxide to simple carboxylic acids. *Organic Letters, 12*, 649–651.

Hermann, B., Blok, K., & Patel, M. K. (2007). Producing bio-based bulk chemicals using industrial biotechnology saves energy and combats climate change. *Environmental Science and Technology, 41*, 7915–7921.

Huber, G. W., & Corma, A. (2007). Synergies between bio-and oil refineries for the production of fuels from biomass. *Angewandte Chemie International Edition, 46*, 7184–7201.

Huo, Z., Hu, M., Zeng, X., Yun, J., & Jin, F. (2012). Catalytic reduction of carbon dioxide into methanol over copper under hydrothermal conditions. *Catalysis Today, 194*, 25–29.

Jang, T. Y., Chung, K. B., Eom, H. R., Noh, D. K., Song, I. K., Yi, J., & Baeck, S.-H. (2011). Catalytic conversion of lactic acid into propylene glycol over various metals supported on silica. *Research on Chemical Intermediates, 37*, 1275–1282.

Jang, H., Kim, S.-H., Lee, D., Shim, S. E., Baeck, S.-H., Kim, B. S., & Chang, T. S. (2013). Hydrogenation of lactic acid to propylene glycol over a carbon-supported ruthenium catalyst. *Journal of Molecular Catalysis a: Chemical, 380*, 57–60.

Jin, F., & Enomoto, H. (2009). Hydrothermal conversion of biomass into value-added products: Technology that mimics nature. *BioResources, 4*, 704–713.

Jin, F., Gao, Y., Jin, Y., Zhang, Y., Cao, J., Wei, Z., & Smith Jr., R. L. (2011). High-yield reduction of carbon dioxide into formic acid by zero-valent metal/metal oxide redox cycles. *Energy & Environmental Science, 4*, 881–884.

Kieboom, A. & Rantwijk, F. V. (1977). *Hydrogenation and hydrogenolysis in synthetic organic chemistry*. Hydrogenation and Hydrogenolysis in Synthetic Organic Chemistry.

Lipinsky, E., & Sinclair, R. (1986). Is lactic acid a commodity chemical. *Chemical Engineering and Processing, 82*, 26–32.

Litchfield. (1996). Microbiological production of Lactic acid. *Advances in Applied Microbiology, 42*, 45–95.

Resende, F. L., & Savage, P. E. (2010). Effect of metals on supercritical water gasification of cellulose and lignin. *Industrial and Engineering Chemistry Research, 49*, 2694–2700.

Saxena, R., Anand, P., Saran, S., Isar, J., & Agarwal, L. (2010). Microbial production and applications of 1, 2-propanediol. *Indian Journal of Microbiology, 50*, 2–11.

Szmant, H. H. (1989). *Organic building blocks of the chemical industry*. Wiley.

Tian, G., Yuan, H., Mu, Y., He, C., & Feng, S. (2007). Hydrothermal reactions from sodium hydrogen carbonate to phenol. *Organic Letters, 9*, 2019–2021.

Wang, Y., Jin, F., Sasaki, M., Wang, F., Jing, Z., & Goto, M. (2013). Selective conversion of glucose into lactic acid and acetic acid with copper oxide under hydrothermal conditions. *AIChE Journal, 59*, 2096–2104.

Wee, Y. J., Kim, J. N., & Ryu, H. W. (2006). Biotechnological production of lactic acid and its recent applications. *Food Technology and Biotechnology, 44*, 163–172.

Xiao, J., Huo, Z., Ren, D., Zhang, S., Luo, J., Yao, G., & Jin, F. (2015). A novel approach for 1,2-propylene glycol production from biomass-derived lactic acid. *Process Biochemistry, 50*, 793–798.

Xu, L., Huo, Z., Fu, J., & Jin, F. (2014). Highly efficient conversion of biomass-derived glycolide to ethylene glycol over CuO in water. *Chemical Communications, 50*, 6009–6012.

Yan, N., & Dyson, P. J. (2013). Transformation of biomass via the selective hydrogenolysis of CO bonds by nanoscale metal catalysts. *Current Opinion in Chemical Engineering, 2*, 178–183.

Yan, Z.-P., Lin, L., & Liu, S. (2009). Synthesis of γ-valerolactone by hydrogenation of biomass-derived levulinic acid over Ru/C catalyst. *Energy & Fuels, 23*, 3853–3858.

Zhang, Z., Jackson, J. E., & Miller, D. J. (2001). Aqueous-phase hydrogenation of lactic acid to propylene glycol. *Applied Catalysis a: General, 219*, 89–98.

Chapter 7
Catalytic Conversion of Ethyl Lactate to 1,2-Propanediol Over CuO

7.1 Introduction

Over the past several decades, consumption of fossil fuels and their limited supply has necessitated a reduction our dependence on petroleum oil. Biomass as a source for producing high value-added chemicals has attracted much attention due to its excellent properties, such as abundant, renewable, and less pollution. A number of efficient methods in the conversion of biomass to chemicals including various alcohols (Xu et al., 2008), hydroxymethylfurfural (van Putten et al., 2013) and dimethylfuran (Román-Leshkov et al., 2007; Zhou et al., 2013) had been reported.

1,2-PDO, a bulk commodity chemical that is mainly used for the production of unsaturated polyester resins (Saxena et al., 2010) and as an antifreeze and de-icing agent and in pharmaceuticals, cosmetics, and foods (Cameron et al., 1998), is industrially produced by the hydration of propylene oxide derived from petrochemical resource (Lipinsky, 1986). With reducing fossil fuel reserves, the development of renewable and green synthetic methods for the 1,2-PDO production is necessary (Clark, 2006; Corma et al., 2007). Recently, the hydrogenation of biomass-derived ethyl lactate to 1,2-PDO using gaseous hydrogen over transition metal catalysts has attracted much interest, which provides an environmental benign alternative to the petroleum-based process for the 1,2-PDO production (Fan et al., 2008; Feng et al., 2011; Han et al., 2012; Huang et al., 2008; Luo et al., 2004). Ethyl lactate can be easily obtained on a large amount (120,000 ton/year) by the fermentation of renewable sources such as agricultural crops and biomass streams. Several groups had disclosed that Ru-based catalysts were effective methods to significantly increase the activity and selectivity for the conversion of ethyl lactate to 1,2-PDO due to their excellent intrinsic activities. For example, RuB/TiO_2 (Fan et al., 2008), RuB/-Al_2O_3 and RuB/SBA-15 (Luo et al., 2004), Ru/SiO_2 (Feng et al., 2011). Li et al. reported vapor-phase hydrogenolysis of ethyl lactate to 1,2-PDO over Co/SiO_2 and Cu/SiO_2, the 1,2-PDO selectivity was in excess of 98% at 90.2% ethyl lactate conversion (Huang et al., 2008). However, these methods suffer from several drawbacks, such as the use of high-pressure gaseous hydrogen which is not easy to obtain at a reduced

Z. Huo, *Diverse Hydrogen Sources for Biomass-derivatives Conversion*,
https://doi.org/10.1007/978-981-99-1673-3_7

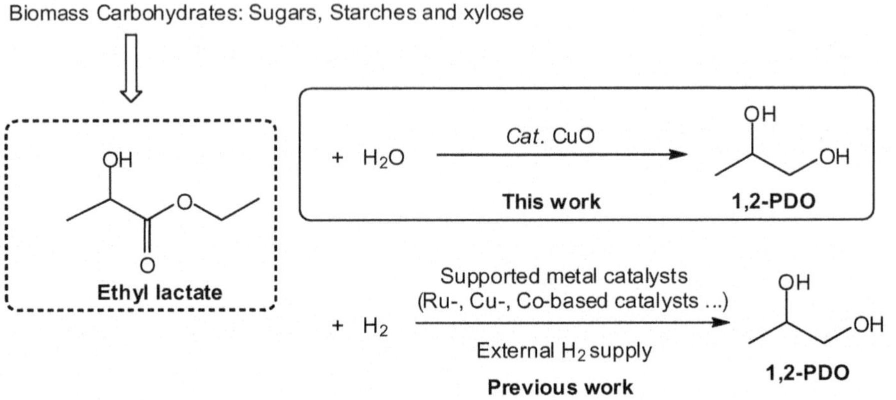

Fig. 7.1 Previous and current process for the formation of 1,2-PDO. Reprinted from Elsevier (Zhang et al., 2016)

energy cost; organic solvents; the use of prepared catalysts, such as Ru-based catalysts supported on some additives (boron, tin or iron, etc.) increased the cost of the preparation and separation process; formation of by-products and longer reaction times. Thus, the development of a cleanly and economically synthetic approach of 1,2-PDO from ethyl lactate is of great importance.

One of the major challenges in the conversion of ethyl lactate to 1,2-PDO is the development of economical hydrogen sources. The previous works showed that hydrothermal process had a promising potential for environmental benign conversion of biomass into useful chemicals (Akiya & Savage, 2002; Kim, 2012; Marshall & Franck, 1981; Watanabe et al., 2004). Water is not only the most abundant hydrogen resource in the process, but also a greener solvent. Recently, we reported an efficient process for the formation of ethylene glycol from glycolide over CuO in water, and high yield of 94% was achieved (Xu et al., 2014). It occurs to us that the conversion of ethyl lactate to 1,2-PDO may be implemented in a similar manner. Herein, we present a highly efficient conversion of ethyl lactate to 1,2-PDO in the presence of CuO in high temperature water (Fig. 7.1).

7.2 Experimental

7.2.1 Experimental Materials

Ethyl lactate (> 97%) as the initial reactant was purchased from TCI. 1,2-PDO ($\geq$ 99.8%, chromatographic grade, J&K Scientific Ltd.) was used for quantitative analysis. Zn (200 mesh, Aladdin) was used as the reductant. Ni, Fe, Al, Mn, Mg, CuO,

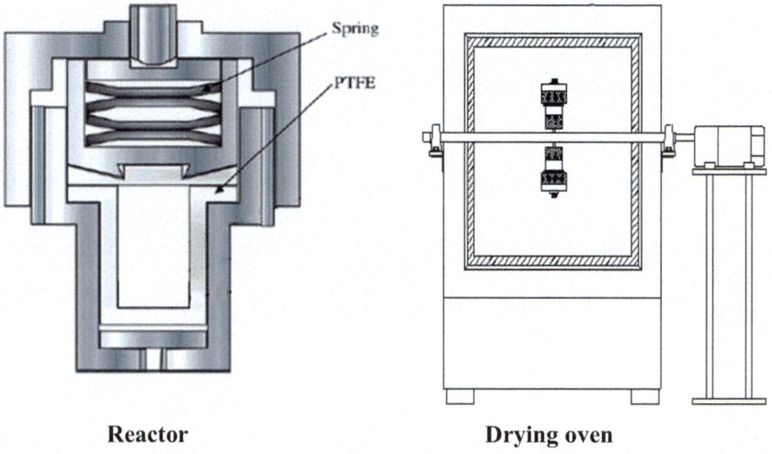

Fig. 7.2 Schematic of batch reactor system. Reprinted from Elsevier (Zhang et al., 2016)

Al_2O_3, CuO, Cu_2O, Fe_2O_3 and Fe_3O_4 (200 mesh, Sinopharm Chemical Reagent Co., Ltd.) were also used in the experiment.

7.2.2 *Experimental Procedure*

All experiments were carried out in a Teflon-lined stainless steel batch reactor with an internal volume of 30 mL. 0.47 mmol ethyl lactate was used in all experiments. The typical procedure for the synthesis of 1,2-PDO was as follows. First, desired ethyl lactate, reductant, catalyst and ultrapure water were loaded into reactor. Then, nitrogen was charged into reactor in order to exclude the effect of air and then the sealed reactor was put into a drying oven, it will take about 20 min to be preheated to the desired temperature. After a desired reaction time, the reactor was quickly moved from the drying oven to cool down. Liquid sample was collected and filtered with 0.45 μm Syringe Filter. Solid sample was collected and washed with deionized water and ethanol several times to remove impurities and dried in the oven at 50 °C for 24 h. The schematic of the reactor system was shown in Fig. 7.2.

7.2.3 *Product Analysis*

After the reaction was quenched, solution samples were collected and filtered for GC–MS analysis (Agilent GC7890A-MS5975C) equipped with an HP Innowax polyethylene glycol capillary column with dimensions of 30 m × 250 μm × 0.25 μm., and HPLC analysis (Agilent 1260 serials, VIS detector). Details on the conditions of

HPLC and GC–MS analyses are available elsewhere (Jin et al., 2001). The total residual organic carbon concentration in liquid samples was also measured with a TOC analyzer (Shimadzu TOC-V). The gas product was detected by thermal conductivity detector (TCD, Agilent Technologies). Thin layer chromatography (TLC) was performed on aluminum-precoated plates of silica gel 60 with an HSGF254 indicator and visualized under UV light or developed by immersion in the solution of 0.6% $KMnO_4$ and 6% K_2CO_3 in water. Solid samples were collected and analyzed by X-ray diffractometer (Shimadzu XRD-6100) to determine the composition and phase purity.

The yield of 1,2-PDO is calculated as the following equation:

$$\text{The yield}(\%) = \frac{\text{the amount of 1,2 - PDO obtained}}{\text{the amount of 1,2 - PDO in theory}} \times 100\%$$

7.3 Results and Discussion

7.3.1 Catalyst Screening

First of all, a series of experiments were carried out to investigate the feasibility of ethyl lactate to 1,2-PDO in the presence of 5 mmol Co and 26 mmol Zn with 17% water filling at 250 °C for 150 min. GC–MS image of the liquid product after the reaction clearly showed that the desired 1,2-PDO was observed (Fig. 7.3). The yield of 37.6% was achieved in the presence of Co (Table 7.1, entry 1). The solid samples after the reaction were analyzed by XRD (Fig. 7.4a), which found that Co still existed in the solid sample, while Zn was oxidized to ZnO, indicating that Zn acted as a reductant and Co acted as a catalyst for the conversion of ethyl lactate to 1,2-PDO. Next, we investigated the performance of other various catalysts, such as CuO, Cu, Fe, MnO_2, ZrO_2, Co_2O_3 and so on, and the results were shown in Table 7.1. The results indicated that CuO, Co and Cu_2O exhibited excellent performance for the conversion of ethyl lactate (entries 1, 11 and 12), and CuO gave much better results than Co and Cu_2O. The yield of 86% was obtained with no surplus initial starting materials observed by thin layer chromatography (TLC) (entry 12). Other catalysts, such as Fe, Cu, Cr, Al_2O_3, Fe_2O_3, Ni_2O_3, Fe_3O_4, MnO_2, Co_2O_3, TiO_2, ZrO_2 and $CuFe_2O_4$ were ineffective, and no desired products were observed (entries 2–10, 13–15). The blank test showed that no desired 1,2-PDO was obtained, indicating the importance of the combined use of both Zn and CuO in the process (entries 16–18). The XRD of solid samples was showed in Fig. 7.4b when CuO was used as a catalyst. It is noteworthy that the Cu rather than CuO was observed in the solid sample, the most likely reason is that CuO was reduced completely to Cu by in situ formed hydrogen by the oxidation of Zn in water (Kubo & Nakagawa, 2012). In order to confirm this assumption, we carried out the experiment under the same reaction

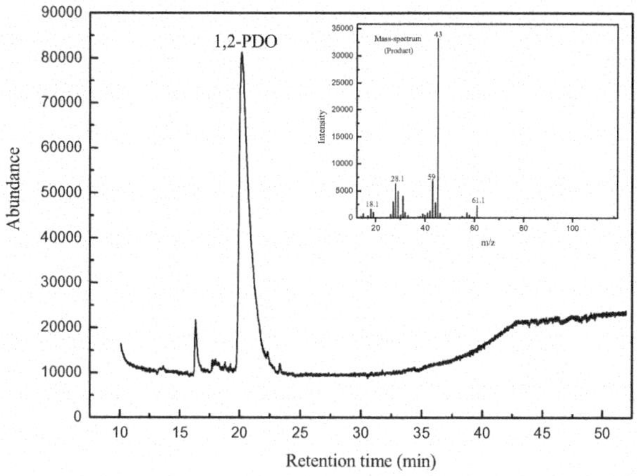

Fig. 7.3 GC–MS spectrum of the liquid sample. Reprinted from Elsevier (Zhang et al., 2016)

conditions without the addition of ethyl lactate. Cu was found from XRD, indicating CuO would be reduced completely to Cu by in situ formed hydrogen during the reaction.

Next, we investigated the effect of different amount of CuO with remaining other conditions unchanged. The results were shown in Fig. 7.5. The yield of 1,2-PDO increased remarkably from 3 to 5 mmol, and then decreased as the amount of CuO was further increased. The decreasing yield may be because more CuO would lead to more cracking products such as propionic acid. It is obvious that 5 mmol CuO is the most suitable amount.

7.3.2 *The Optimization of the Various Parameters*

To investigate the effect of reductant on the yield of 1,2-PDO, the reactions of ethyl lactate were carried out in the presence of CuO at 250 °C for 150 min. Table 7.2 showed that the desired 1,2-PDO was yielded when Zn and Fe were used, while the other reductants, such as Mg, Ni, Mn, Al were ineffective and no desired product was obtained at all. Zn is the most efficient reductant as the yield of 1,2-PDO is 86%. Thus, Zn was chosen as a reductant in the following experiments. Next, the amount of Zn was optimized in the presence of CuO at 250 °C for 150 min. As shown in Fig. 7.6, the results indicated that the most suitable amount is 26 mmol. But the yield of 1,2-PDO decreased significantly when the amount of Zn exceeded 26 mmol, which demonstrated that too much hydrogen pressure may have negative effects in the conversion of ethyl lactate to 1,2-PDO.

The change of Zn oxidation to ZnO during the reaction at different time was monitored by XRD. As shown in Fig. 7.4b, peaks of Zn gradually reduced until it disappeared, whereas peaks of ZnO increased gradually as time increased. However, peak of Cu was also observed, indicating that CuO was reduced gradually as Zn

Table 7.1 Effect of catalysts on the yield of 1,2-PDO

Entry	Reductant	Catalyst	1,2-PDO	Lactic acid (%)	Acetic acid (%)	Propionic acid (%)	Hydroxy-acetone (%)
1	Zn	Co	37.6	7.0	8.7	9.0	7.0
2	Zn	Fe	0	6.3	8.0	12.0	4.5
3	Zn	Cu	0	6.0	9.7	10.2	1.0
4	Zn	Cr	0	16	7.3	6.0	5.4
5	Zn	Al_2O_3	0	5.5	11.0	12.4	6.3
6	Zn	Fe_2O_3	0	6.1	10.0	13.0	1.5
7	Zn	Ni_2O_3	0	5.5	12.2	13.0	4.9
8	Zn	Fe_3O_4	0	5.7	9.8	14.3	7.4
9	Zn	MnO_2	0	8.3	8.9	15.6	5.6
10	Zn	TiO_2	0	7.0	10.9	15.3	3.5
11	Zn	Cu_2O	66.5	8.0	12.3	11.7	6.0
12	Zn	CuO	86.0	5.6	10.0	5.3	6.0
13	Zn	ZrO_2	0	7.9	15.0	14.0	3.6
14	Zn	Co_2O_3	0	5.6	12.1	13.0	4.0
15	Zn	$CuFe_2O_4$	0	5.6	11.1	13.0	1.0
16	Zn	–	0	44.7	9.0	7.0	1.0
17	–	CuO	0	5.3	10.2	5.3	1.4
18	–	–	0	13.2	7.8	10.0	2.1

Reaction conditions: Zn: 26 mmol; catalyst: 5 mmol; time: 150 min; temp: 250 °C; water filling: 17%. Reprinted from Elsevier (Zhang et al., 2016)

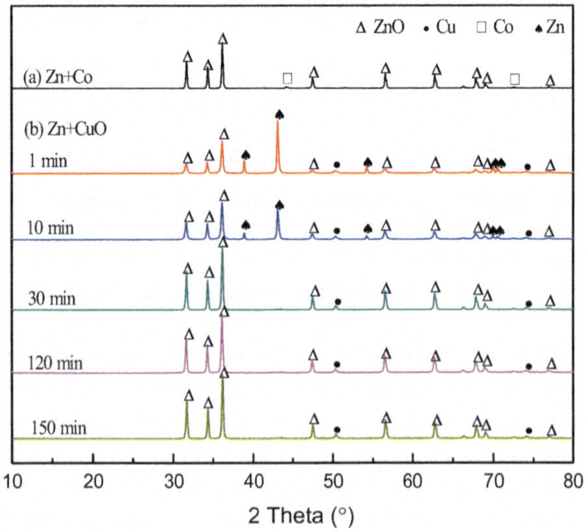

Fig. 7.4 XRD patterns of the solid samples Reprinted from Elsevier (Zhang et al., 2016)

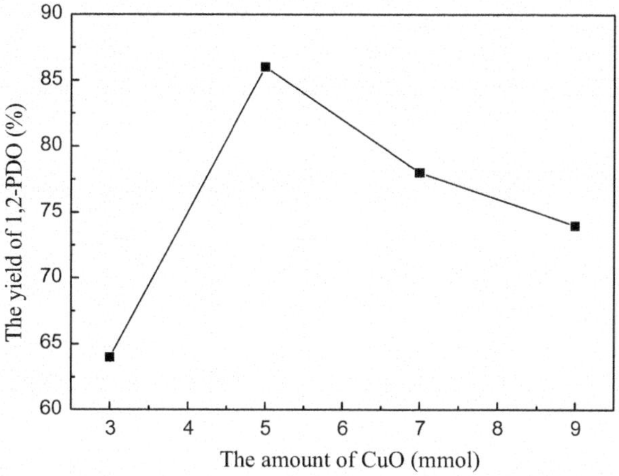

Fig. 7.5 Effect of CuO amount (Zn: 26 mmol; time: 150 min; temp: 250 °C; water filling: 17%). Reprinted from Elsevier (Zhang et al., 2016)

Table 7.2 Effect of reductants on the yield of 1,2-PDO

Entry	Reductant	1,2-PDO	Lactic acid (%)	Acetic acid (%)	Propionic acid (%)	Hydroxy-Acetone (%)
1	Zn	86.0	13.2	7.8	10.0	2.1
2	Mg	0	6.0	19.6	9.2	1.0
3	Ni	0	6.3	8.0	12.0	4.5
4	Mn	0	7.0	18.7	9.0	5.6
5	Fe	37.8	7.3	8.7	8.5	7.6
6	Al	0	6.5	8.6	12.4	3.2

Reaction conditions: reductant: 26 mmol; CuO: 5 mmol; water filling: 17%; time: 150 min; temp: 250 °C). Reprinted from Elsevier (Zhang et al., 2016)

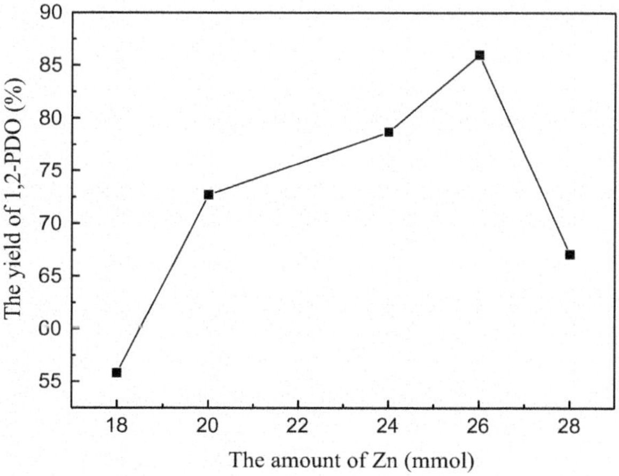

Fig. 7.6 Effect of Zn amount (CuO: 5 mmol; time: 150 min; temp: 250 °C; water filling: 17%). Reprinted from Elsevier (Zhang et al., 2016)

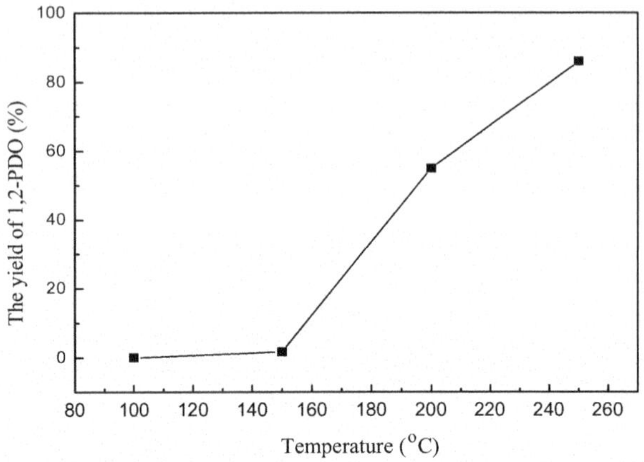

Fig. 7.7 Effect of the temperature (Zn: 26 mmol; CuO: 5 mmol; time: 150 min; water filling: 17%). Reprinted from Elsevier (Zhang et al., 2016)

was oxidized to ZnO. These results indicate that ethyl lactate can be converted to 1,2-PDO over CuO by in situ formed hydrogen by the oxidation of Zn in water.

The effect of temperatures from 100 to 250 °C on the yields of 1,2-PDO was investigated. As shown in Fig. 7.7, the reaction did not take place at 100 °C and no desired 1,2-PDO was observed. The yield of 1,2-PDO improved gradually as the reaction temperature increased gradually from 100 to 150 °C, and then remarkably increased from 150 to 250 °C. The results indicated high temperature contributed to the conversion of ethyl lactate to 1,2-PDO. However, the temperature was limited to 250 °C due to the limiting temperature of the Teflon material.

Experiments were performed by changing the reaction time from 1 to 3 h at 250 °C to examine the effect of reaction time on the yield of 1,2-PDO. Figure 7.8 demonstrated that the yield of 1,2-PDO increased as reaction time increased from 1 h to 2.5 h, and maximum yield was reached when reaction time was 2.5 h. However, the yield of 1,2-PDO decreased after 2.5 h. This phenomenon may be attributed to the decomposition of produced 1,2-PDO during the reaction. In order to confirm this assumption, an experiment was designed as follows, 1,2-PDO acted as the initial reactant in the presence of 5 mmol Cu and 26 mmol ZnO with 17% water filling at 250 °C for 2 h. The HPLC indicated that the initial 120 mmol/L concentration of 1,2-PDO was decreased to 100 mmol/L, and XRD analysis showed that Cu and ZnO had not changed. It is obvious that a long reaction time has negative effect on the yield of 1,2-PDO obtained in the presence of Cu and ZnO during the reaction (Larcher & Patrice, 2000).

Further study was to optimize suitable water filling from 10 to 45%. Figure 7.9 indicates that the 1,2-PDO yield first increased significantly when the water filling was changed from 10 to 17%. The yield of 1,2-PDO increased slightly as the water filling increased from 17 to 42% and then decreased after 42%. The optimum water filling was 42% to obtain the highest yield of 93.6%. The decreasing yield at a higher water filling is most likely due to decomposition of obtained 1,2-PDO with the increase of pressure when water filling increased.

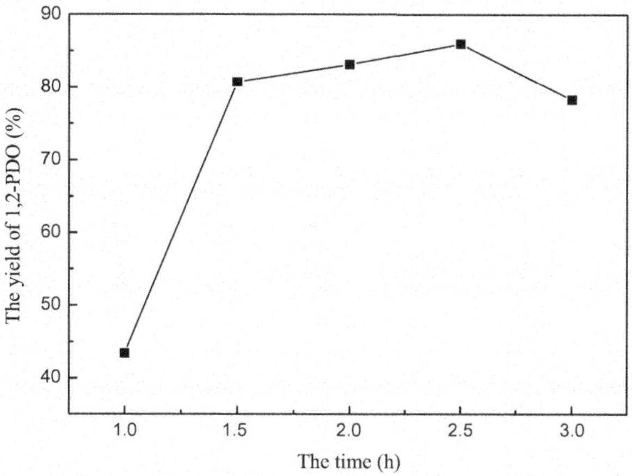

Fig. 7.8 Effect of the time (Zn: 26 mmol; CuO: 5 mmol; temp: 250 °C; water filling: 17%). Reprinted from Elsevier (Zhang et al., 2016)

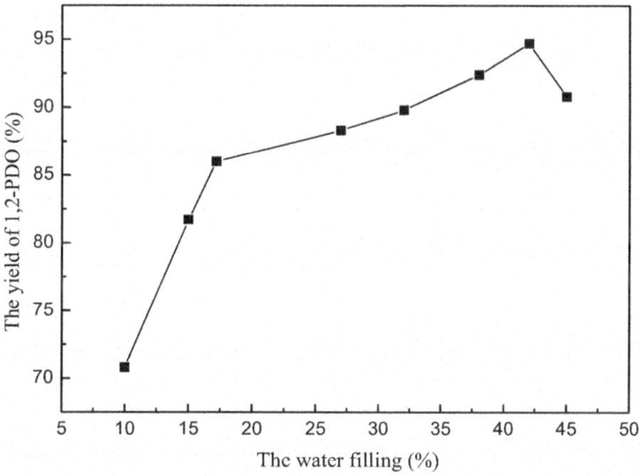

Fig. 7.9 Effect of water filling (Zn: 26 mmol; CuO: 5 mmol; time: 150 min; temp: 250 °C). Reprinted from Elsevier (Zhang et al., 2016)

Finally, a carbon balance was conducted at the optimized condition by TOC and found that approximately 97.8% ethyl lactate was converted into liquid products. And only H_2 as the gaseous product was detected.

7.3.3 The Formation of 1,2-PDO from Ethyl Lactate Using Gaseous Hydrogen

The catalytic hydrogenation of ethyl lactate using gaseous hydrogen instead of in situ hydrogen to 1,2-PDO was also investigated. The reaction was carried out in the presence of 5 mmol CuO, 2 MPa H_2, 42% water filling at 250 °C for 2.5 h. Unfortunately, lactic acid rather than 1,2-PDO was obtained, indicating gases hydrogen was ineffective for the 1,2-PDO production from ethyl lactate (Eq. 7.1).

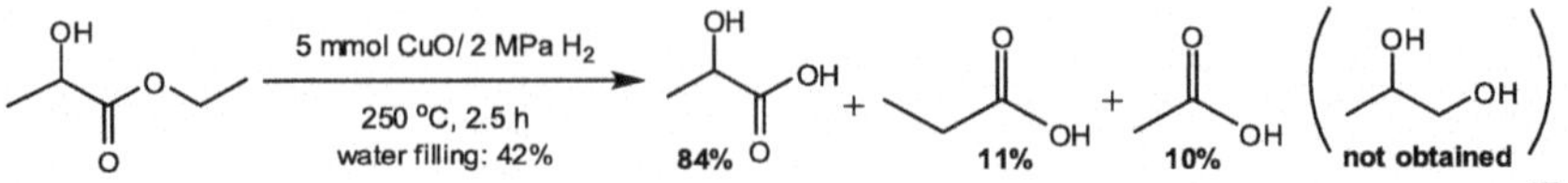

(7.1)

7.3.4 Mechanistic Studies

Based on the previous and current results, the pathway for the conversion of ethyl lactate to 1,2-PDO over CuO was proposed in Fig. 7.10. First, hydrogen was formed via the oxidation of Zn in water (Elliott, 2008; Resende & Savage, 2010). The in situ-formed hydrogen was adsorbed on the surface of Cu/CuO/ZnO though the formation of hydrogen bonds. Then, ethyl lactate was also adsorbed on the surface of Cu/CuO/ZnO by the combination of Cu/CuO/ZnO with carbonylic C and O atoms (Kieboom & van Rantwijk, 1977; Yan et al., 2009). When the hydrogen reacted with ethyl lactate molecules, an intermediate 2-hydroxypropanal was obtained. Furthermore, both the intermediate 2-hydroxypropanal and the hydrogen formed were adsorbed on the surface of Cu/CuO/ZnO, H was added to C=O and an intermediate linked by an α-bond was formed. Finally, the intermediate acquired one more H from the surface of Cu/CuO/ZnO and the desired 1,2-PDO was obtained.

To investigate the possibility of experimental pathway for the transformation, validation test for the conversion of intermediate 2-hydroxypropanal to 1,2-PDO was conducted. As a result, intermediate 2-hydroxypropanal was converted completely to 1,2-PDO in 90% yield under optimum conditions (Eq. 7.2). This validation process illustrates that the proposed pathway is acceptable.

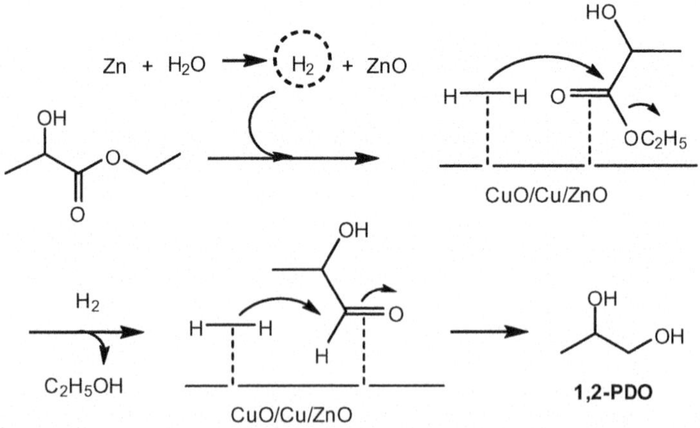

Fig. 7.10 Proposed pathway for the formation of 1,2-PDO. Reprinted from Elsevier (Zhang et al., 2016)

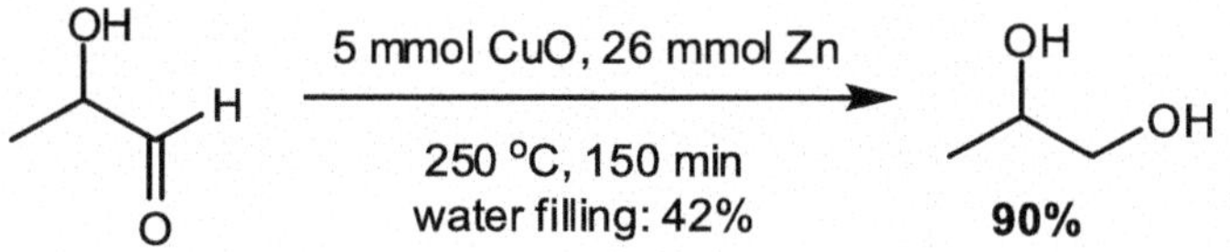

(7.2)

7.4 Conclusions

We developed a highly efficient conversion of biomass-derived ethyl lactate to 1,2-PDO over CuO in high temperature water. CuO catalyst exhibits high catalytic activity for the transformation. A high yield of 93.6% was achieved in the presence of 26 mmol Zn and 5 mmol CuO with 42% water filling at 250 °C for 150 min. The use of gaseous hydrogen was ineffective and no desired 1,2-PDO was obtained. This study provided a highly efficient process for the conversion of biomass-derived ethyl lactate to 1,2-PDO. Further study will be taken to develop efficient methods for the conversion of biomass-derived compounds to value-added chemicals.

References

Akiya, N., & Savage, P. E. (2002). Roles of water for chemical reactions in high-temperature water. *Chemical Reviews, 102*, 2725–2750.

Cameron, D. C., Altaras, N. E., Hoffman, M. L., & Shaw, A. J. (1998). Metabolic engineering of propanediol pathways. *Biotechnology Progress, 14*, 116–125.

Clark, J. H. (2006). Green chemistry: Today (and tomorrow). *Green Chemistry, 8*, 17–21.

Corma, A., Iborra, S., & Velty, A. (2007). Chemical routes for the transformation of biomass into chemicals. *Chemical Reviews, 107*, 2411–2502.

Elliott, D. C. (2008). Catalytic hydrothermal gasification of biomass. *Biofuels, Bioproducts and Biorefining, 2*, 254–265.

Fan, G.-Y., Zhang, Y., Zhou, Y.-F., Li, R.-X., Chen, H., & Li, X.-J. (2008). Efficient hydrogenation of ethyl lactate To 1,2-propanediol over Ru–B/TiO_2 in aqueous solution. *Chemistry Letters, 37*, 852–853.

Feng, J., Xiong, W., Jia, Y., Wang, J., Liu, D., & Qin, R. (2011). Hydrogenation of ethyl lactate over ruthenium catalysts in an additive-free catalytic system. *Reaction Kinetics, Mechanisms and Catalysis, 104*, 89–97.

Han, Z., Rong, L., Wu, J., Zhang, L., Wang, Z., & Ding, K. (2012). Catalytic hydrogenation of cyclic carbonates: A practical approach from CO_2 and epoxides to methanol and diols. *Angewandte Chemie International Edition, 51*, 13041–13045.

Huang, L., Zhu, Y., Zheng, H., Du, M., & Li, Y. (2008). Vapor-phase hydrogenolysis of biomass-derived lactate to 1,2-propanediol over supported metal catalysts. *Applied Catalysis a: General, 349*, 204–211.

Jin, F.-M., Kishita, A., Moriya, T., & Enomoto, H. (2001). Kinetics of oxidation of food wastes with H_2O_2 in supercritical water. *Journal of Supercritical Fluids, 19*, 251–262.

Kieboom, A. P. G., & Van Rantwijk, F. (1977). *Hydrogenation and hydrogenolysis in synthetic organic chemistry*. Springer.

Kim, A. S. (2012). Dissipative hydrodynamics of rigid spherical particles. *Chemistry Letters, 41*, 1128–1130.
Kubo, S., & Nakagawa, M. (2012). Gram-scale synthesis of zinc oxide nanorods in basic ethanol solutions. *Chemistry Letters, 41*, 1137–1138.
Larcher, D., & Patrice, R. (2000). Preparation of metallic powders and alloys in polyol media: A thermodynamic approach. *Journal of Solid State Chemistry, 154*, 405–411.
Lipinsky, E. S. (1986). *State of the art in bioconversion*. Elsevier.
Luo, G., Yan, S., Qiao, M., Zhuang, J., & Fan, K. (2004). Effect of tin on Ru-B/Γ-Al_2o_3 catalyst for the hydrogenation of ethyl lactate to 1,2-propanediol. *Applied Catalysis a: General, 275*, 95–102.
Marshall, W. L., & Franck, E. U. (1981). Ion product of water substance, 0–1000 °C, 1–10,000 bars new international formulation and its background. *Journal of Physical and Chemical Reference Data, 10*, 295–304.
Resende, F. L. P., & Savage, P. E. (2010). Effect of metals on supercritical water gasification of cellulose and lignin. *Industrial and Engineering Chemistry Research, 49*, 2694–2700.
Román-Leshkov, Y., Barrett, C. J., Liu, Z. Y., & Dumesic, J. A. (2007). Production of dimethylfuran for liquid fuels from biomass-derived carbohydrates. *Nature, 447*, 982–985.
Saxena, R. K., Anand, P., Saran, S., Isar, J., & Agarwal, L. (2010). Microbial production and applications of 1,2-propanediol. *Indian Journal of Microbiology, 50*, 2–11.
Van Putten, R.-J., Van Der Waal, J. C., De Jong, E., Rasrendra, C. B., Heeres, H. J., & De Vries, J. G. (2013). Hydroxymethylfurfural, a versatile platform chemical made from renewable resources. *Chemical Reviews, 113*, 1499–1597.
Watanabe, M., Sato, T., Inomata, H., Smith, R. L., Arai, K., Kruse, A., & Dinjus, E. (2004). Chemical reactions of C_1 compounds in near-critical and supercritical water. *Chemical Reviews, 104*, 5803–5822.
Xu, G., Shao, C., & Xiu, Z. (2008). Optimizing control of bio-dissimilation process of glycerol to 1,3-propanediol. *Chinese Journal of Chemical Engineering, 16*, 128–134.
Xu, L., Huo, Z., Fu, J., & Jin, F. (2014). Highly efficient conversion of biomass-derived glycolide to ethylene glycol over CuO in water. *Chemical Communications, 50*, 6009–6012.
Yan, Z.-P., Lin, L., & Liu, S. (2009). Synthesis Of Γ-Valerolactone by hydrogenation of biomass-derived Levulinic acid over Ru/C catalyst. *Energy & Fuels, 23*, 3853–3858.
Zhang, S., Huo, Z. B., Ren, D. Z., Luo, J., Fu, J., Li, L., & Jin, F. M. (2016). Catalytic conversion of ethyl lactate to 1,2-propanediol over CuO. *Chinese Journal of Chemical Engineering, 24*, 126–131.
Zhou, C., Yu, X., Ma, H., He, R., & Vittayapadung, S. (2013). Optimization on the conversion of bamboo shoot shell to levulinic acid with environmentally Benign acidic ionic liquid and response surface analysis. *Chinese Journal of Chemical Engineering, 21*, 544–550.

Chapter 8
Highly Selective Hydrothermal Production of Cyclohexanol from Biomass-Derived Cyclohexanone Over Cu Powder

8.1 Introduction

Cyclohexanol has been considered as very important intermediate and is widely used in the synthesis of hexane diacid and hexanolactam. Moreover, cyclohexanol could also be used in pharmaceuticals, plasticisers, surfactant, paint, industry solvent and so on. Due to its importance, much attention has been paid to developing efficient methods for the production of cyclohexanol in recent years. Among these processes, catalytic hydrogenation of cyclohexanone to cyclohexanol is an attractive route since the reaction proceeds under mild conditions.

We all know that the hydrogenation of ketones to alcohols is an important reaction in organic chemistry (Gorgas et al., 2014; Mahdi & Stephan, 2014). Cyclohexanone is a significant by-product in the degradation process of lignocellulose. Lignocellulose is an important biomass which makes up a high proportion of biomass (An et al., 2014). Besides, biomass and biomass-derived compounds as a promising renewable resource instead of fossil fuel is attracting more and more attention because they can provides not only clean energy but also raw material for chemical industry (Cheng & Huber, 2011; Raju et al., 2015; Singh et al., 2010; Westerhof et al., 2011). Therefore, the conversion of biomass-derived cyclohexanone to produce cyclohexanol will be more meaningful.

Currently, catalytic hydrogenation of cyclohexanone to cyclohexanol attracts more and more attention due to its promising use in the field of green chemistry (Bhar & Guha, 2004; Campbell et al., 2001; Fujihara et al., 2010). Bhar et al. reported the hydrogenation of ketones to the corresponding alcohols by aluminum powder (5 eq.) in the presence of sodium hydroxide in $MeOH:H_2O = 2:1$, the yield of cyclohexanol was only 19% (Bhar & Guha, 2004). Campbell et al. used Meerwein-Pondorf-Verley reduction by aluminum complexes in toluene at room temperature, after 2 h, the yield of cyclohexanol reached 96% (Campbell et al., 2001). Fujihara et al. used copper complexes as catalyst and Ph_2SiH_2 as hydrogen source in toluene at room temperature, the catalytic activity was good. However, they used complicated and toxic complexes, and the hydrogen source was expensive (Singh et al., 2010).

Z. Huo, *Diverse Hydrogen Sources for Biomass-derivatives Conversion*,
https://doi.org/10.1007/978-981-99-1673-3_8

Other noble metal, such as Ir (Wei et al., 2013), Rh (Niyomura et al., 2005), Ru (Spasyuk et al., 2013), can also showed high catalytic activity in the conversion of cyclohexanone and obtained good results. However, these previous works still have some obvious defects, such as expensive prepared noble metal catalysts (Niyomura et al., 2005; Spasyuk et al., 2013; Wei et al., 2013), organic solvents (Bhar & Guha, 2004; Campbell et al., 2001; Fujihara et al., 2010; Niyomura et al., 2005; Spasyuk et al., 2013) and longer reaction times (6–12 h) (Wei et al., 2013). Therefore, one of the great challenges is to develop a useful conversion process of cyclohexanone in the presence of high-efficiency commercial catalysts. As a solvent, water has some merits that other organic solvent doesn't have, such as cheap, non-toxic, non-flammable, non-explosive and environmental friendly (Watson, 2012). If water could be widely used as a solvent for organic reaction, it would bring about enormous achievement in environmental, economic and safety aspects (Wei et al., 2013).

In previous works, we and other groups had developed many methods to convert biomass and biomass-derived compounds to value-added chemicals in high-temperature water (Huo et al., 2015; Sudasinghe et al., 2014; Xu et al., 2014). For example, we developed an efficient process for the selective conversion of glucose to lactic acid under alkaline hydrothermal conditions with the highest yield of 25% (Huo et al., 2014). Recently, we reported a method for the formation of ethylene glycol from glycolide over CuO in water, and high yield of 94% was achieved (Xu et al., 2014). Also we also developed catalytic conversion of ethyl lactate to 1,2-propanediol in the presence of CuO with the yield of 93% (Xiao et al., 2015).

Based on the studies above, we herein present a highly efficient hydrothermal conversion of biomass-derived cyclohexanone to cyclohexanol with high yield and high selectivity using in situ-formed hydrogen in water in the presence of Cu catalyst. A mechanism of cyclohexanone conversion was also proposed (Eq. 8.1).

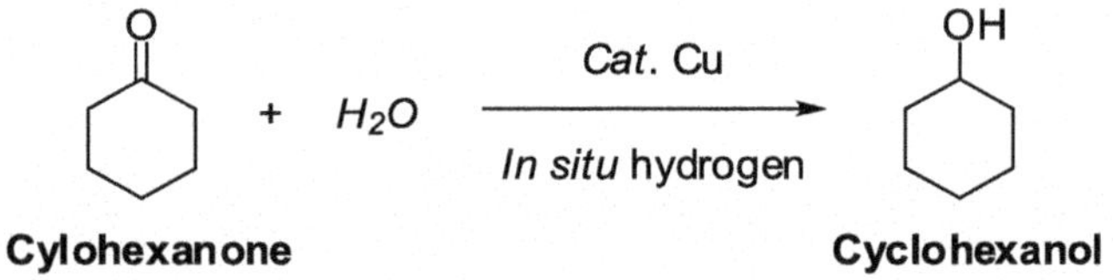

(8.1)

8.2 Experimental

8.2.1 Experimental Materials

Cyclohexanone (≥ 99.5%) as the initial reactants was purchased from J&K Scientific Ltd. Cyclohexanol (≥ 98.5%) was used for quantitative analysis. Zn (200 mesh, Sinopharm Chemical Reagent Co., Ltd.) was used as the reductant. Fe, Fe_2O_3, Fe_3O_4, Cu, CuO, Cu_2O, Ni, NiO, Co, SiO_2, ZrO_2, Al_2O_3, TiO_2, Pd/C, active carbon (AR,

200mesh, Sinopharm Chemical Reagent Co., Ltd.) and Mn (AR, 200 mesh, Aladdin) and Sn (AR, 200 mesh, Sigma-Aldrich) were also used in the experiments. And two catalysts, $Cu(OH)PO_4$ and $CuFe_2O_4$,were prepared by ourselves according to the recent publication (Baghbanian & Farhang, 2014).

8.2.2 Experimental Procedure

All the experiments were performed in a Teflon-lined steel batch reactor whose internal volume is 30 mL. Cyclohexanone (0.193 mmol, 20 μL) was used for all experiments in this study. The procedure for the synthesis of cyclohexanol was as follows. First of all, desired cyclohexanone, deionized water, reactant and catalyst were loaded into reactor. Then, after filling in with nitrogen to remove air, the reactor was putted into a drying oven. After a certain period of time, we took out the reactor from drying oven immediately. Due to the oven setting, it would take 20 min to be preheated to the desired temperature, so we prolonged 20 min to reaction times. The reactor was cooled by an air fan. Finally, liquid sample was collected and filtered by 0.45 μm Syringe Filter. Solid sample was collected and washed by deionized water for several times and dried in drying oven at 60 °C for 24 h. The samples would be analysed later. The schematic diagram of reactor and drying oven was shown in Fig. 8.1.

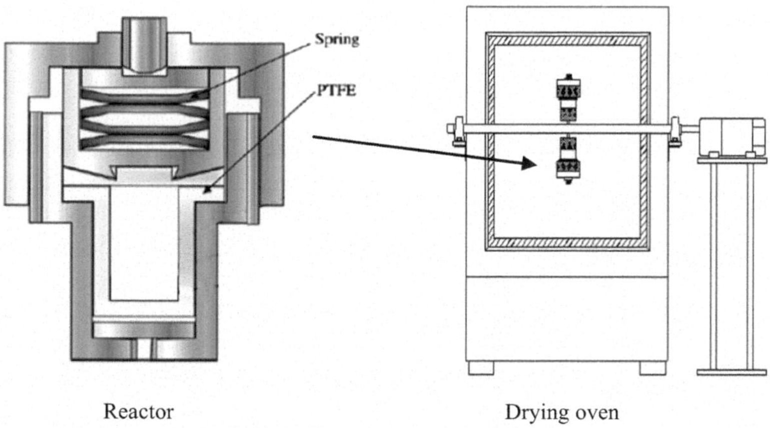

Fig. 8.1 Schematic of batch reactor system. Reprinted from Royal Society of Chemistry. Reprinted from Catalysis Today (Song et al., 2016)

8.2.3 *Product Analysis*

After the reaction, liquid samples were confirmed with GC–MS (Agilent 7890A GC system, 5975C inert MSD with Triple-Axis Detector) equipped with an HP Innowax polyethylene glycol capillary column (30 m × 250 μm × 0.5 μm). The qualification of cyclohexanol was confirmed with GC-FID (Agilent 7890A GC system, 5975C inert MSD with Triple-Axis Detector) equipped with an HP Innowax polyethylene glycol capillary column (30 m × 250 μm × 0.5 μm). The solid samples were analysed with X-ray diffraction (Shimadzu, Lab-XRD-6100). The total organic carbon concentration was confirmed by a TOC Analyzer (Shimadzu TOC-V).

The yield of cyclohexanol was calculated on the basis of the following equation:

$$\text{The yield } (\%) = \frac{\text{The more of cyclohexanol obtained}}{\text{The more of cyclohexanol}} \times 100\%$$

8.3 Results and Discussion

8.3.1 *Catalyst Screening*

Initially, a series of experiments were carried out to investigate the feasibility of converting cyclohexanone to cyclohexanol in the presence of 5 mmol metal or metal oxides and 25 mmol Zn with 25% water filling (7.5 mL H_2O) at 250 °C for 150 min. The results are summarized in Table 8.1. The conversion of cyclohexanone was higher and cyclohexanol yield of 92.2% was achieved in the presence of both Cu and Zn (entry 1). It is obvious that the unique peak of desired cyclohexanol and 100.1 of its mass-to-charge ratio (m/z) can be observed by GC–MS spectrum as shown in Fig. 8.2a. From Fig. 8.2b, the peak of cyclohexanone is getting smaller and smaller until disappearing completely while the peak of cyclohexanol becomes bigger gradually as the time increased from 0 to 150 min, indicating that cyclohexanone was converted completely. Various Cu complexes, such as CuO, Cu_2O, $Cu_2(OH)PO_4$ and $CuFe_2O_4$, were investigated and gave desired products in moderate to high yields, but the yields were not superior to that obtained with Cu catalyst (entries 2–5). Co, Sn, Ni and NiO were also effective for the conversion of cyclohexanone to afford good yields (entries 6–9). However, other metals or metal oxides, such as Fe, Fe_2O_3, Pd/C, Mn, active carbon and so on, proved to be completely ineffective and lower yields were obtained (entries 10–19). The reaction of cyclohexanone afforded only 11.5% yield or no desired cyclohexanol was observed at all in the absence of either Cu or Zn or both, indicating that the importance of combination of Cu and Zn (entries 20–22).

As illustrated in Fig. 8.3, the composition of the solid samples after the reaction was determined by XRD patterns to investigate the role of metals. Zn was found to

Table 8.1 Catalyst screening on the yield of cyclohexanol[a]

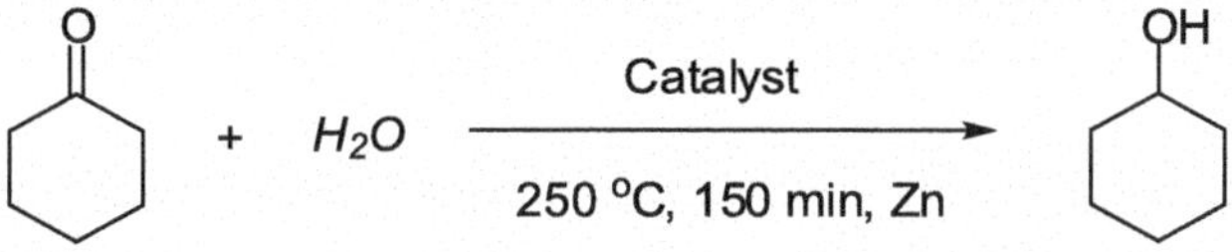

Entry	Reductant	Catalyst	Cyclohexanol (%)
1	Zn	Cu	92.2
2	Zn	Cu_2O	71.2
3	Zn	CuO	70.5
4	Zn	$Cu_2(OH)PO_4$	47.4
5	Zn	$CuFe_2O_4$	57.6
6	Zn	Co	86.7
7	Zn	Sn	64.0
8[b]	Zn	Ni	70.8
9	Zn	NiO	75.4
10	Zn	Fe	18.2
11	Zn	Fe_2O_3	15.4
12	Zn	Fe_3O_4	5.3
13	Zn	SiO_2	8.7
14	Zn	ZrO_2	14.4
15	Zn	Al_2O_3	17.8
16	Zn	TiO_2	11.1
17	Zn	Pd/C	5.0
18	Zn	Mn	16.2
19	Zn	Active carbon	9.7
20	Zn	–	11.5
21	–	Cu	0
22	–	–	0

[a]Reaction conditions: [cyclohexanone (0.193 mmol, 20 μL), 25 mmol Zn, 5 mmol catalyst, 25% water filling (7.5 mL H_2O), 250 °C, 150 min]. [b]Byproduct phenol of 17.0% was obtained with the conversion of 87.8%. Reprinted from Catalysis Today (Song et al., 2016).

be oxidized to ZnO while Cu or Co, Sn, Ni, Fe kept unchanged and still existed in the solid samples. This suggested that in this process Zn acted as a reductant while Cu or Co, Sn, Ni, Fe acted as catalysts. Mn was oxidized to Mn_3O_4 after reaction and the XRD patterns was shown in Fig. 8.4.

Next, the reactions were conducted to investigate the effect of the amount of Cu catalyst on the yield of cyclohexanol. In Fig. 8.8, the result indicated the yield of cyclohexanol increased significantly with the amount of Cu increased from 0 to 8 mmol, and the maximum value of 96.4% was obtained at 7.5 mmol. When the

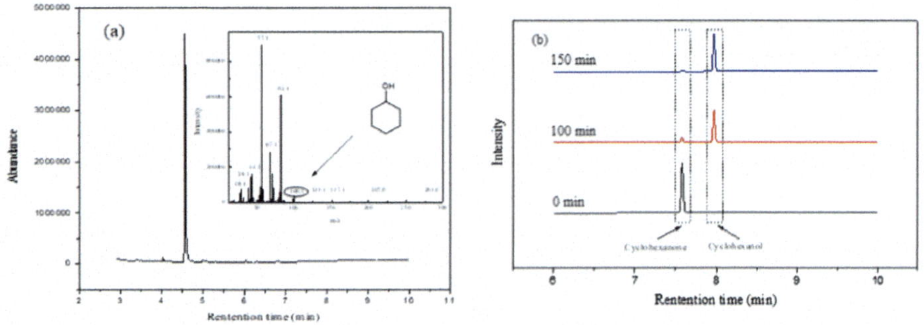

Fig. 8.2 **a** GC–MS spectrum of the liquid samples; **b** GC-FID of liquid samples at different reaction time. Reprinted from Catalysis Today (Song et al., 2016)

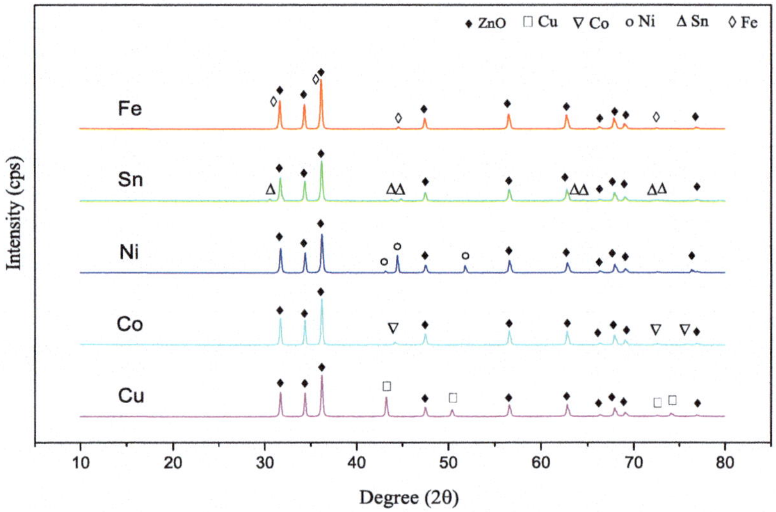

Fig. 8.3 XRD patterns of the solid samples. Reprinted from Catalysis Today (Song et al., 2016)

amount of Cu powder was more than 7.5 mmol, the yield of cyclohexanol decreased gradually as the amount of Cu increased. From the result we can reasonably infer that excess Cu can cause the reversible conversion of formed cyclohexanol to cyclohexanone leading to the decreasing yield. Thus, 7.5 mmol Cu catalyst was the most suitable for the conversion of cyclohexanone.

The Cu catalyst was collected to observe the change before and after the reaction. After the reaction, Zn was removed by using HCl. The Cu powder was filtered, collected and washed by deionized water, then dried using an oven at 60 °C for 24 h. As was shown in Fig. 8.5, we can see that there has a big change in shape of Cu powder after the reaction, it becomes rough and irregular.

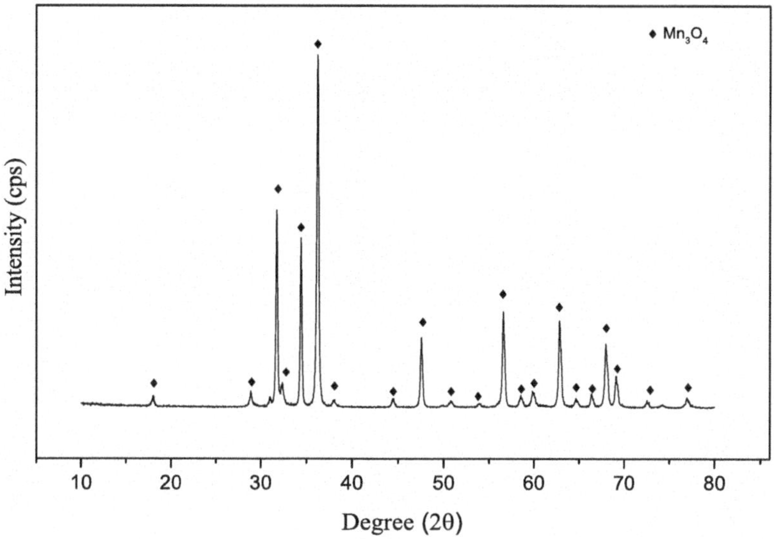

Fig. 8.4 XRD patterns of Mn after the reaction [cyclohexanone (0.193 mmol, 20 μL), 25 mmol Zn, 5 mmol Mn, 25% water filling (7.5 mL H_2O), 250 °C, 150 min]. Reprinted from Catalysis Today (Song et al., 2016)

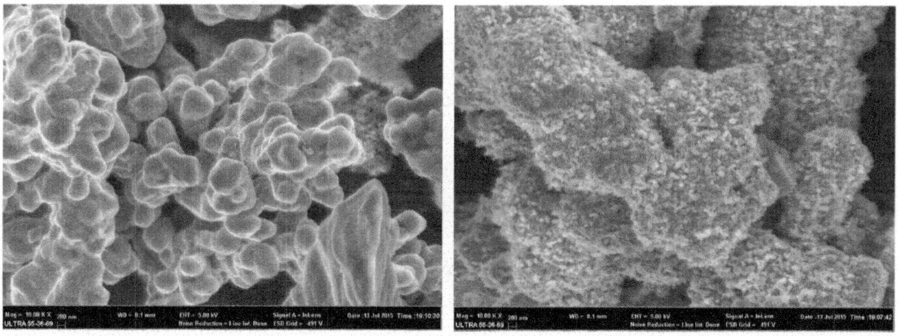

Fig. 8.5 Scanning electron microscopy (SEM) images of Cu powder. Left: before the reaction; Right: after the reaction. Reprinted from Catalysis Today (Song et al., 2016)

We then examined the reaction to confirm whether or not Cu catalyst can cause the reversible conversion of the formed cyclohexanol. The formed cyclohexanol as starting material was reacted for 150 min in the presence of 7.5 mmol Cu and 20 mmol ZnO obtained after the reaction at 250 °C, and the 18.6% yield of cyclohexanone was obtained and the 39.7% yield of cyclohexanol was recovered. This result clearly indicated that this reaction was reversible over Cu catalyst (Eq. 8.2). As illustrated in Fig. 8.6, gaseous phase CO and H_2 were observed by GC–MS and TCD.

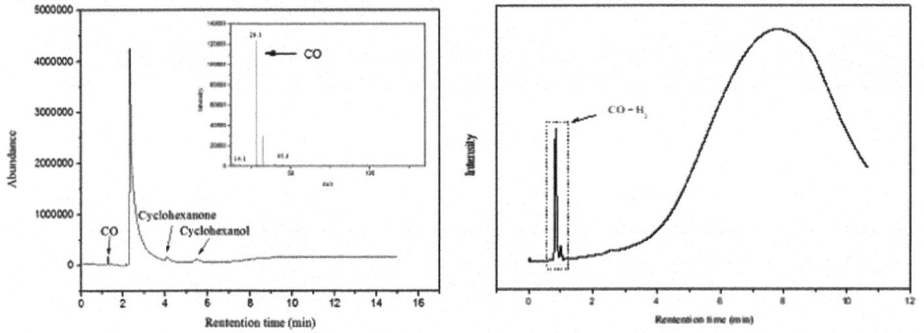

Fig. 8.6 GC–MS and TCD of gaseous products [cyclohexanone (0.193 mmol, 20 μL), 20 mmol ZnO, 7.5 mmol Cu, 25% water filling (7.5 mL H_2O), 250 °C, 150 min]. Reprinted from Catalysis Today (Song et al., 2016)

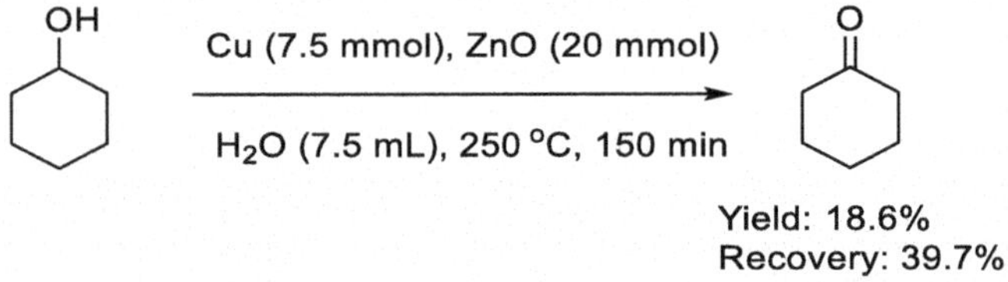

(8.2)

8.3.2 *Optimization of Active Metals*

After that, we further studied the effect of various active metals as reductants on the yield of cyclohexanol when other reaction conditions unchanged [cyclohexanone (0.193 mmol, 20 μL), 25 mmol reductant, 7.5 mmol Cu, 25% water filling (7.5 mL H_2O), 250 °C, 150 min]. The details are showed in Table 8.2 and Fig. 8.7. When active Mg, Al, Ni, Fe were chosen to be reductants, it can be seen clearly that the efficiency of transformation was as follows, Zn > Mg > Al > Ni > Fe, and the yields were 96.4%, 50.9%, 50.8%, 13.8% and 8.4%, respectively. Obviously, Zn was the most suitable reductant to be used in the process. Then, the amount of Zn was optimized as shown in Fig. 8.8, the yield of product significant increased with increasing amount of Zn. When the amount of Zn was 20 mmol, cyclohexanone was completely converted and the highest yield reached 100%. The decreasing yield might be due to the reversible conversion of obtained cyclohexanol by more hydrogen by the oxidation of excess Zn in water or by the effect of Cu catalyst and ZnO as mentioned above.

Table 8.2 Effect of reductants on the yield of cyclohexanol

Entry	Reductant	Cyclohexanol (%)
1	–	0
2	Zn	96.4
3	Mg	50.9
4	Al	50.8
5	Ni	13.8
6	Fe	8.4

Reaction conditions: [cyclohexanone (0.193 mmol, 20 μL), 25 mmol reductant, 7.5 mmol Cu, 25% water filling (7.5 mL H_2O), 250 °C, 150 min]. Reprinted from Catalysis Today (Song et al., 2016)

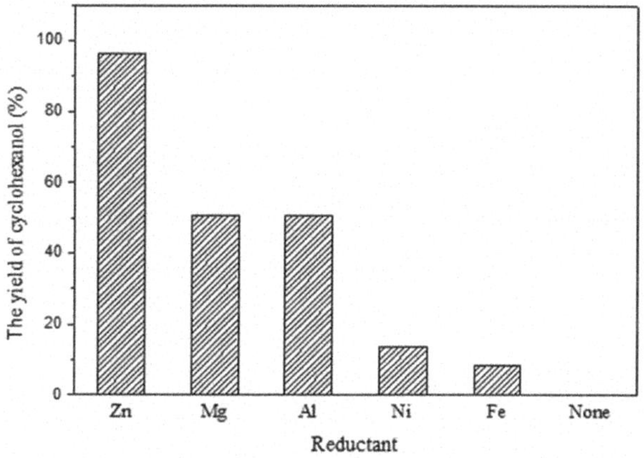

Fig. 8.7 Effect of reductants [cyclohexanone (0.193 mmol, 20 μL), 25 mmol reductant, 7.5 mmol Cu, 25% water filling (7.5 mL H_2O), 250 °C, 150 min]. Reprinted from Catalysis Today (Song et al., 2016)

8.3.3 *Effects of the Various Parameters*

Water filling is also a very influential factor for the conversion of cyclohexanone. As shown in Fig. 8.9, water filling from 15 to 45% has a significant effect for the production of cyclohexanol from cyclohexanone. Considering the security reasons of the Teflon material, water filling must be less than 45% because excessive pressure would cause potential accident during the reaction. Figure 8.9 showed that the yield increased significantly with water filling continuously increasing from 15 to 25%. The optimum water filling was 25% (about 3.87 MPa) to obtain the maximum value of 100%, and then decreased after 25%. This can be speculated that the incremental water pressure is better for promoting the reversible conversion of obtained cyclohexanol when water filling increases, which leads to low yield.

As a general rule, temperature plays an important role in chemical reaction. In the process the increase of the reaction temperature from 100 °C to 250 °C gave higher conversion of cyclohexanone. Figure 8.10 showed that the yield of cyclohexanol increased drastically from 100 °C to 195 °C. After 195 °C, the yield increased

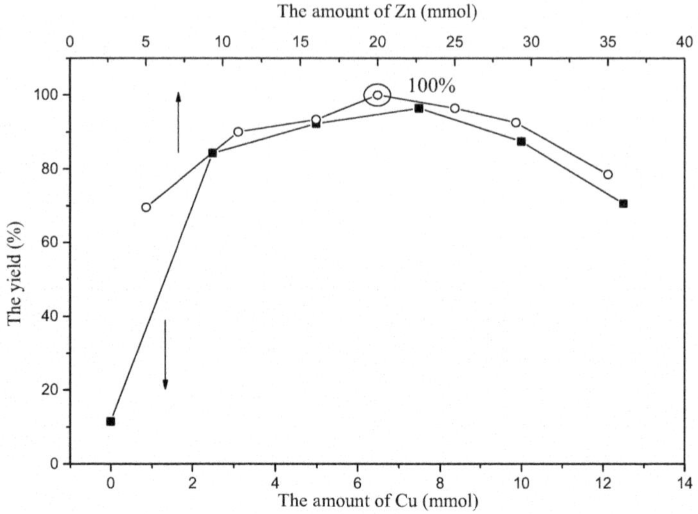

Fig. 8.8 **a** Effect of Cu amount [cyclohexanone(0.193 mmol, 20 μL), 25 mmol Zn, 25% water filling (7.5 mL H_2O), 250 °C, 150 min]; **b** Effect of Zn amount [cyclohexanone (0.193 mmol, 20 μL), 7.5 mmol Cu, 25% water filling (7.5 mL H_2O), 250 °C, 150 min]. Reprinted from Catalysis Today (Song et al., 2016)

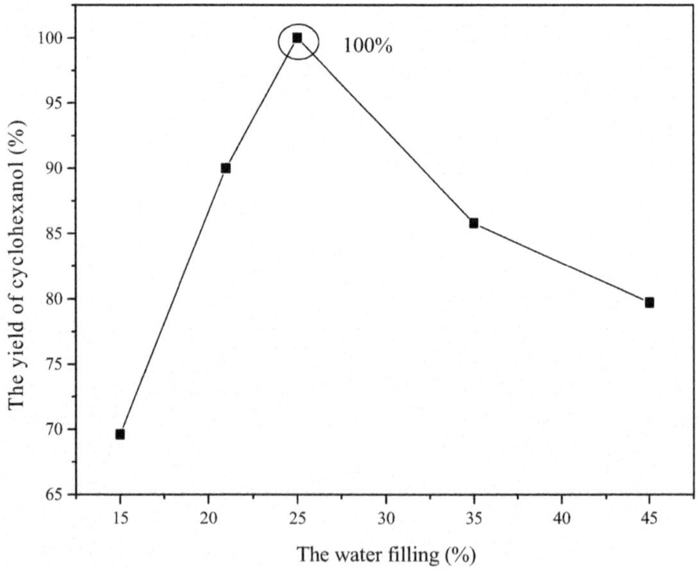

Fig. 8.9 Effect of water filling [cyclohexanone (0.193 mmol, 20 μL), 20 mmol Zn, 7.5 mmol Cu, 250 °C, 150 min]. Reprinted from Catalysis Today (Song et al., 2016)

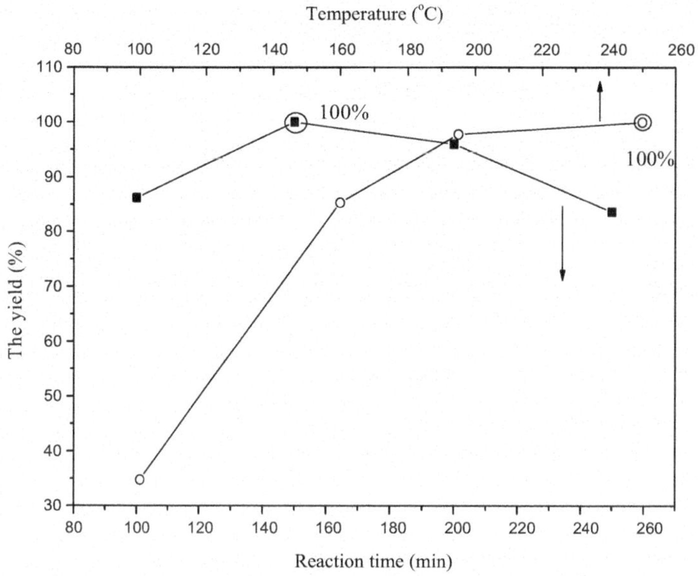

Fig. 8.10 **a** Effect of time [cyclohexanone (0.193 mmol 20 μL), 20 mmol Zn, 7.5 mmol Cu, 25% water filling (7.5 mL), 250 °C]; **b** effect of temperature [cyclohexanone (0.193 mmol 20 μL), 20 mmol Zn, 7.5 mmol Cu, 25% water filling (7.5 mL H_2O), 150 min]. Reprinted from Catalysis Today (Song et al., 2016)

gradually. At 250 °C, cyclohexanone was completely converted and a maximum value of 100% was reached, it clearly indicated that higher reaction temperature are more favourable to the production of cyclohexanol from cyclohexanone. However, it is worth noting that because the maximum safety temperature of the Teflon material of reactor was 250 °C, higher than 250 °C was not optimized.

Having optimized 250 °C as the best reaction temperature for the production of cyclohexanol, further experiments were carried out to investigate the effect of reaction time from 100 to 250 min on the yield of cyclohexanol. In Fig. 8.10, it can be seen that the yield of cyclohexanol improved drastically as the reaction time increased from 100 to 150 min and the yield of 100% was obtained at 150 min. When the time had increased more than 150 min, a decrease in the conversion of cyclohexanone as well as a decreasing yield of cyclohexanol was observed. Evidential experiment had found that obtained cyclohexanol as starting material could be converted reversibly in the presence of Cu and ZnO obtained by the oxidation of Zn at 250 °C for 150 min as mentioned above. The result indicated that the longer reaction time was unfavourable for the production of cyclohexanol.

Finally, we carried out the experiments to investigate carbon balance by TOC and TCD, and found that 100% cyclohexanone was converted into cyclohexanol under the optimal reaction conditions. Only gaseous H_2 was detected according to the result of TCD.

8.3.4 Cyclohexanone Hydrogenation by Added Hydrogen

To investigate the conversion of cyclohexanone using added gaseous hydrogen as a hydrogen source in the presence of 7.5 mmol Cu powder with 3 MPa H_2 at 250 °C for 150 min, the reaction also proceeded well. The experiments were performed in a SUS 316 stainless steel reactor whose internal volumes is 42 mL with a gas collection valve. First, the desired cyclohexanone, Cu powder and water was loaded into the reactor. Then the reactor was sealed by a gas collection valve. The desired amount of H_2 was calculated through the ideal gas law, equals to the H_2 produced by the oxidation of Zn in water. After that, the reactor was moved into the induction furnace reactor. The temperature procedure was as follows. The reactor was heated from room temperature to 250 °C in 15 min, when the reaction time was over, the reactor was cooled from 250 °C to room temperature by natural cooling. At the end, the gaseous and liquid samples were collected, the gaseous products were detected by TCD, and the liquid sample was collected and filtered with 0.45 μm Syringe Filter and detected by GC-FID. As expected, the yield of 75.1% was obtained although it is much lower than that of in situ-formed hydrogen (Eq. 8.3).

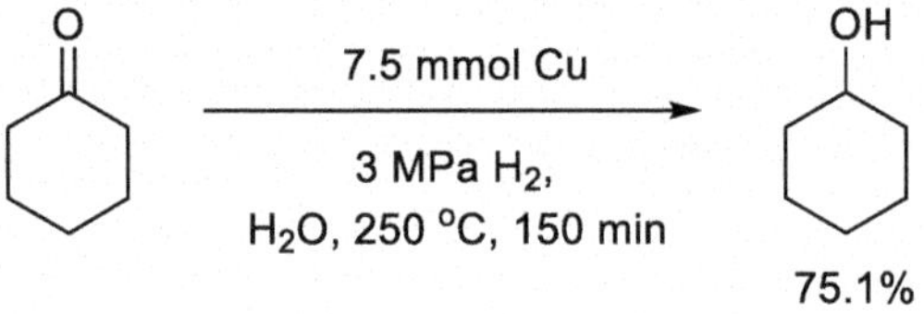

(8.3)

8.3.5 Plausible Mechanism

Based on the experimental results above, a plausible reaction mechanism for the conversion of cyclohexanone to cyclohexanol over Cu powder was proposed in Fig. 8.11. First, hydrogen in situ-formed by the oxidation of Zn in water was adsorbed on the surface of Cu powder. At the same time, cyclohexanone was also absorbed on the surface of Cu/ZnO with carbonylic C and O atoms. After that, the high chemical activity hydrogen atom attacked cyclohexanone molecules, hydrogen atom was added to carbonylic C atom. Then, another hydrogen atom from the surface of Cu/ZnO was attached to O atom, which leaded to the formation of hydroxyl. Finally, the desired cyclohexanol was obtained.

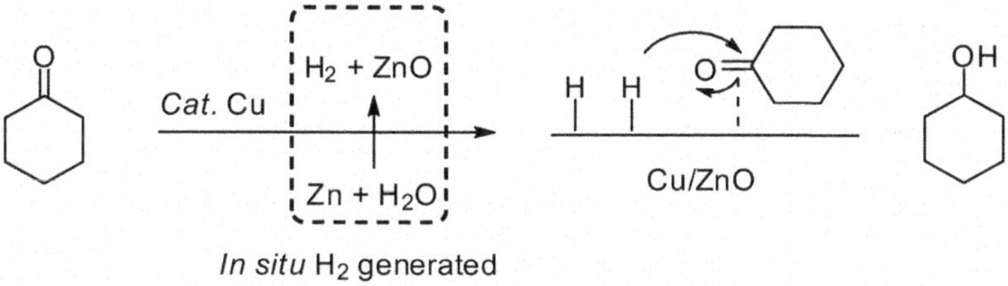

Fig. 8.11 Proposed mechanism for the formation of cyclohexanol. Reprinted from Catalysis Today (Song et al., 2016)

8.4 Conclusions

In summary, we explored a highly efficient synthesis of cyclohexanol from biomass-derived cyclohexanone over Cu powder. As a result, with optimized conditions in hand, cyclohexanone was completely converted and obtained a maximum value of 100% in the presence of 20 mmol Zn and 7.5 mmol Cu powder with 25% water filling at 250 °C for 150 min, which was higher than that of 75.1% yield using pure H_2. From both academic and industrial viewpoints, the present study is very meaningful to provide an efficient process for the selective conversion of biomass-derived cyclohexanone to cyclohexanol with high yield.

References

An, Y.-X., Li, N., Zong, M.-H., & Lou, W.-Y. (2014). Easily measurable pH as an indicator of the effectiveness of the aqueous cholinium ionic liquid-based pretreatment of lignocellulose. *RSC Advances, 4*, 55635–55639.

Baghbanian, S. M., & Farhang, M. (2014). $CuFe_2O_4$ nanoparticles: A magnetically recoverable and reusable catalyst for the synthesis of quinoline and quinazoline derivatives in aqueous media. *RSC Advances, 4*, 11624–11633.

Bhar, S., & Guha, S. (2004). The remarkable role of water during the chemoselective reduction of ketones mediated by metallic aluminium. *Tetrahedron, 45*, 3775–3777.

Campbell, E. J., Zhou, H., & Nguyen, S. T. (2001). Catalytic Meerwein–Pondorf–Verley reduction by simple aluminum complexes. *Organic Letters, 3*, 2391–2393.

Cheng, Y.-T., & Huber, G. W. (2011). Chemistry of furan conversion into aromatics and olefins over HZSM-5: A model biomass conversion reaction. *ACS Catalysis, 1*, 611–628.

Fujihara, T., Semba, K., Terao, J., & Tsuji, Y. (2010). Copper-catalyzed hydrosilylation with a bowl-shaped phosphane ligand: Preferential reduction of a bulky ketone in the presence of an aldehyde. *Angewandte Chemie International Edition, 49*, 1472–1476.

Gorgas, N., Stöger, B., Veiros, L. F., Pittenauer, E., Allmaier, G., & Kirchner, K. (2014). Efficient hydrogenation of ketones and aldehydes catalyzed by well-defined iron(II) PNP pincer complexes: Evidence for an insertion mechanism. *Organometallics, 33*, 6905–6914.

Huo, Z., Fang, Y., Ren, D., Zhang, S., Yao, G., Zeng, X., & Jin, F. (2014). Selective conversion of glucose into lactic acid with transition metal ions in diluted aqueous NAOH solution. *ACS Sustainable Chemistry & Engineering, 2*, 2765–2771.

Huo, Z.-B., Liu, J.-K., Yao, G.-D., Zeng, X., Luo, J., & Jin, F.-M. (2015). Efficient hydrothermal conversion of cellulose into methane over porous Ni catalyst. *Applied Catalysis a: General, 490*, 36–41.

Mahdi, T., & Stephan, D. W. (2014). Enabling catalytic ketone hydrogenation by frustrated Lewis Pairs. *Journal of the American Chemical Society, 136*, 15809–15812.

Niyomura, O., Iwasawa, T., Sawada, N., Tokunaga, M., Obora, Y., & Tsuji, Y. (2005). A bowl-shaped phosphine as a ligand in rhodium-catalyzed hydrosilylation: Rate enhancement by a Mono(phosphine) rhodium species. *Organometallic, 24*, 3468–3475.

Raju, S., Moret, M.-E., & Klein Gebbink, R. J. M. (2015). Rhenium-catalyzed dehydration and deoxydehydration of alcohols and polyols: Opportunities for the formation of olefins from biomass. *ACS Catalysis, 5*, 281–300.

Singh, N. R., Delgass, W. N., Ribeiro, F. H., & Agrawal, R. (2010). Estimation of liquid fuel yields from biomass. *Environmental Science and Technology, 44*, 5298–5305.

Song, Z., Ren, D., Wang, T., Jin, F., Jiang, Q., & Huo, Z. (2016). Highly selective hydrothermal production of cyclohexanol from biomass-derived cyclohexanone over Cu powder. *Catalysis Today, 274*, 94–98.

Spasyuk, D., Smith, S., & Gusev, D. G. (2013). Replacing phosphorus with sulfur for the efficient hydrogenation of esters. *Angewandte Chemie International Edition, 52*, 2538–2542.

Sudasinghe, N., Cort, J. R., Hallen, R., Olarte, M., Schmidt, A., & Schaub, T. (2014). Hydrothermal liquefaction oil and hydrotreated product from pine feedstock characterized by heteronuclear two-dimensional NMR spectroscopy and FT-ICR mass spectrometry. *Fuel, 137*, 60–69.

Watson, W. J. W. (2012). How do the fine chemical, pharmaceutical, and related industries approach green chemistry and sustainability? *Green Chemistry, 14*, 251–259.

Wei, Y., Xue, D., Lei, Q., Wang, C., & Xiao, J. (2013). Cyclometalated iridium complexes for transfer hydrogenation of carbonyl groups in water. *Green Chemistry, 15*, 629–634.

Westerhof, R. J. M., Brilman, D. W. F., Garcia-Perez, M., Wang, Z., Oudenhoven, S. R. G., Van Swaaij, W. P. M., & Kersten, S. R. A. (2011). Fractional condensation of biomass pyrolysis vapors. *Energy & Fuels, 25*, 1817–1829.

Xiao, J., Huo, Z., Ren, D., Zhang, S., Luo, J., Yao, G., & Jin, F. (2015). A novel approach for 1, 2-propylene glycol production from biomass-derived lactic acid. *Process Biochemistry, 50*, 793–798.

Xu, L., Huo, Z., Fu, J., & Jin, F. (2014). Highly efficient conversion of biomass-derived glycolide to ethylene glycol over CuO in water. *Chemical Communications, 50*, 6009–6012.

Chapter 9
Efficient Conversion of Dimethyl Phthalate to Phthalide Over CuO in Aqueous Media

9.1 Introduction

During the past several decades, the energy crises have led many countries to seek renewable energy resources to reduce the over dependence on fossil fuels (Huber et al., 2006; Jin & Enomoto, 2011). After years of over-consumption on unsustainable petroleum, the develop renewable and "green" sources of energy to reduce this use of fossil fuels and reduce CO_2 emissions has become imperative (Chheda et al., 2007; Corma et al., 2007; Hermann et al., 2007; Huber & Corma, 2007). Biomass is a kind of abundant, renewable and relatively non-polluted resource, these qualities make it an excellent candidate for the production of high value-added chemicals. For this reason, the utilization of biomass currently is receiving increasing attention.

PHT has wide applications in medicine, antiseptics and organic synthesis. It is an intermediate chemical with important uses in, for example, the treatment of circulatory and heart diseases. PHT can also act as a versatile building block in organic synthesis, especially, the PHT is useful in the synthesis of functionalized naphthalenes, anthracenes and naphthacene natural products (Donner, 2013; Hernández et al., 2007; Mitchell & Russell, 1995; Rathwell & Brimble, 2007; Snieckus, 1990). Due to its beneficial industrial applications, much attention has been paid to the synthesis of PHT. Currently, synthesis of PHT typically involves an industrial process from petroleum-derived phthalimide. DMP, an important biomass compound, can be easily obtained from the plentiful biomass sources, says the celery (Hayashi et al., 1967). Using renewable biomass compound DMP to obtain PHT is an efficient method which has the potential for the future commercializaton. However, to date, very few studies have focused on this field. Franco Piacenti et al. reported the hydrogenation of DMP into the corresponding PHT using a ruthenium hydride complex $H_4Ru_4(CO)_8(PBu_3)_4$ (Matteoli et al., 1984). Toshiharu Yokayama et al. discussed turning DMP into PHT with a yield of 55% over the ZrO_2 and Cr-ZrO_2 catalyst at a temperature of over 350 °C (Yokoyama & Fujita, 2004). Matthew L. Clarke et al. used $LiHBEt_3$ as a catalyst for the reaction and achieved a yield of 34% (Clarke et al., 2007). However, challenges still lurk in these methods from both academia

Z. Huo, *Diverse Hydrogen Sources for Biomass-derivatives Conversion*,
https://doi.org/10.1007/978-981-99-1673-3_9

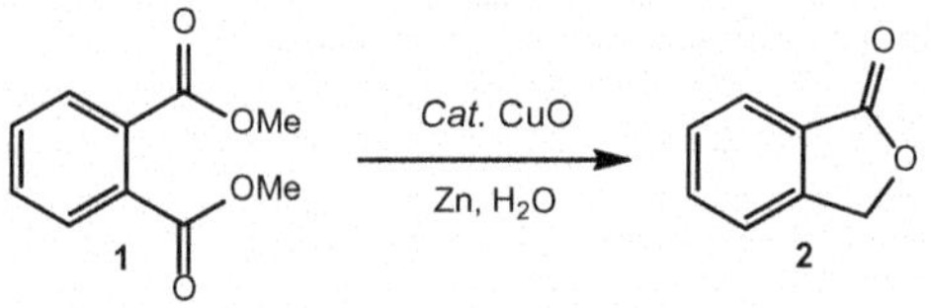

Fig. 9.1 Conversion of DMP to PHT in the presence of CuO. Reprinted from Elsevier (Fu et al., 2019)

and industry perspective. For example, the high-pressure gaseous hydrogen, which is oftenly used, is difficult to store and is sometimes unstable; the PHT yield is low; the use of expensive synthesized catalysts and organic solvents are not economic and could be detrimental to the environment; and plus, a longer reaction time. Therefore, the development of an efficient and scalable process to produce PHT from DMP is very important.

Previously, our group and others have reported on a series of experiments concentrated on the conversion of biomass into value-added chemicals using a process involving high-temperature water (Chheda et al., 2007; Tian et al., 2007; He et al., 2010; Akiya & Savage, 2002; Duan & Savage, 2011; Fang et al., 2008; Wang et al., 2013; Jin & Heiji, 2009; Huo et al., 2012; Jin et al., 2011; Huber et al., 2004; Huber et al., 2005). In this process, water is not only an environmentally friendly reaction medium, and it is also an abundant hydrogen resource provided by the oxidation of active metal in water (Kruse & Dinjus, 2007). In addition, we have also reported a highly-efficient production of diols from biomass-derived glycolide or lactic acid over CuO in a high yield with high selectivity (Xiao et al., 2015; Xu et al., 2014). On the basis of these findings, if we can extend this concept to the conversion of DMP, PHT may be produced in a similar manner. Herein, we present a highly efficient conversion of DMP to PHT in the presence of CuO in aqueous media (Fig. 9.1).

9.2 Experimental

9.2.1 Experimental Materials

DMP (> 99%) as the initial reactant was purchased from J&K. PHT (≥ 99%, chromatographic grade, J&K Scientific Ltd.) was used for quantitative analysis. Zn (200 mesh, Aladdin) was used as the reductant. Ni, Fe, Al, Mn, Mg, CuO, Al_2O_3, CuO, Cu_2O, Fe_2O_3 and Fe_3O_4 (200 mesh, Sinopharm Chemical Reagent Co., Ltd) were also used in the experiment.

9.2.2 *Experimental Procedure*

All experiments were carried out in a Tefion-lined stainless steel batch reactor with an internal volume of 30 mL. 0.5 mmol DMP was used in all experiments. The typical procedure for the synthesis of PHT was as follows. First, desired DMP, reductant, catalyst and ultrapure water were loaded into the reactor. Then, the nitrogen was charged into the reactor in order to exclude the effect of air and then the sealed reactor was put into a drying oven, it will take about 20 min to be preheated to the desired temperature. After a desired reaction time, the reactor was quickly moved out from the drying oven to cool down. Liquid sample was collected and extracted by 10 mL ethyl acetate, then filtered with 0.45 μm Syringe Filter. Solid sample was collected and washed with deionized water and ethanol several times to remove impurities and dried in the oven at 50 °C for 24 h. The schematic of the reactor system was shown in Fig. 9.2.

The experiments were conducted in a SUS 316 stainless steel reactor whose internal volumes is 42 mL with a gas collection valve. First, the desired DMP, CuO and water was loaded into the reactor. Then the reactor was sealed by a gas collection valve. The desired amount of H_2 was calculated through the ideal gas law, equals to the H_2 produced by the oxidation of Zn in water. After that, the reactor was moved into the induction furnace reactor. The temperature procedure was as follows. The reactor was heated from room temperature to 250 °C in 15 min, when the reaction time was over, the reactor was cooled from 250 °C to room temperature by natural cooling. Finally, the gaseous and liquid samples were collected, the gaseous products were detected by TCD, and the liquid sample was collected and filtered with 0.45 μm Syringe Filter and detected by GC-FID.

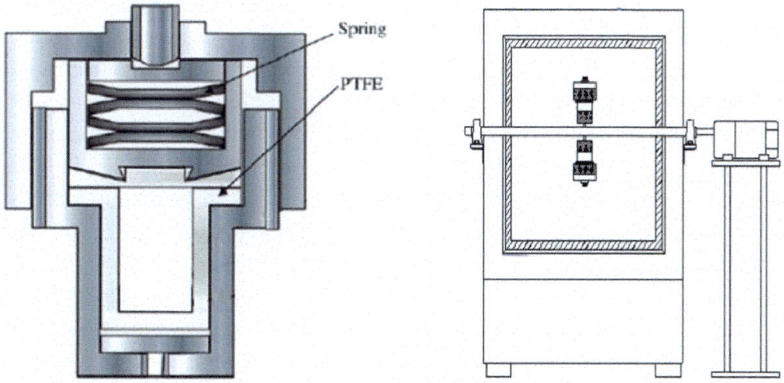

Reactor Drying oven

Fig. 9.2 Schematic of batch reactor system. Reprinted from Elsevier (Fu et al., 2019)

9.2.3 *Product Analysis*

After the reaction, the solution samples were collected and filtered for GC–MS analysis (Aglient GC7890A-MS5975C) equipped with an HP-5 polyethylene glycol capillary column with dimensions of 30 m × 250 μm × 0.25 μm., and GC-FID. Details on the conditions of GC–MS/FID analyses are available elsewhere (Clarke et al., 2007). The total residual organic carbon concentration in liquid samples was also measured with a TOC analyzer (Shimadzu TOC-V). The gas product was detected by thermal conductivity detector (TCD, Agilent Technologies). Thin layer chromatography (TLC) was performed on aluminum-precoated plates of silica gel 60 with an HSGF254 indicator and visualized under UV light or developed by immersion in the solution of 0.6% $KMnO_4$ and 6% K_2CO_3 in water. Solid samples were collected and analyzed by X-ray diffractometer (Shimadzu XRD-6100) to determine the composition and phase purity. The yield of PHT is calculated as the following equation:

$$\text{The yield } (\%) = \frac{\text{the amount of PHT obtained}}{\text{the amount of PHT in theory}} \times 100\%$$

9.3 Results and Discussion

9.3.1 *Catalyst Screening*

Initially, we investigate the influence of metal or metal oxides on the production of DMP to PHT. The reaction of DMP was conducted in the presence of 5 mmol Cu and 25 mmol Zn at 250 °C for 120 min. The result showed that the reaction proceeded efficiently and produced the desired PHT at a 61% yield, as determined by GC-FID NMR and GC–MS (Figs. 9.3, and 9.4). Next, we screened a series of metal or metal oxides to determine the most suitable one. The results are summarised in Table 9.1. The reaction did not proceed at all in the absence of both reductant and catalyst or in the presence of the catalysts only (entries 1, 3–4). The reaction with 25 mmol Zn but without the addition of any catalyst, gave the desired PHT at a 24% yield (entry 2), this result could be reckoned as an evidence of the importance of the combined use of the reductant and catalyst. Among the various catalysts we investigated (entries 5–17), the use of CuO improved the transformation efficiency and afforded PHT at a better yield (entries 8). However, the use of Fe_3O_4 is completely ineffective (entry 11). $CuFe_2O_4$ and Ni gave lower yield (entries 15, 17).

As the XRD patterns of solid residues after the reaction shown in Fig. 9.5a, the Cu still existes while Zn has been oxidized into ZnO, it is obviously that the Cu acted as a catalyst and Zn acted as a reductant in the reaction process. According to the result of Fig. 9.5b, unexpectedly, the Cu instead of CuO is observed in the

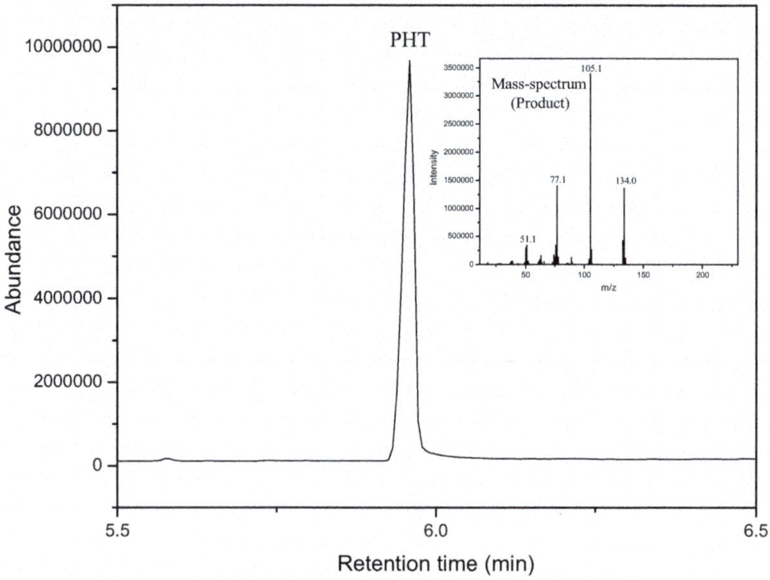

Fig. 9.3 GC–MS spectrum of PHT. Reprinted from Elsevier (Fu et al., 2019)

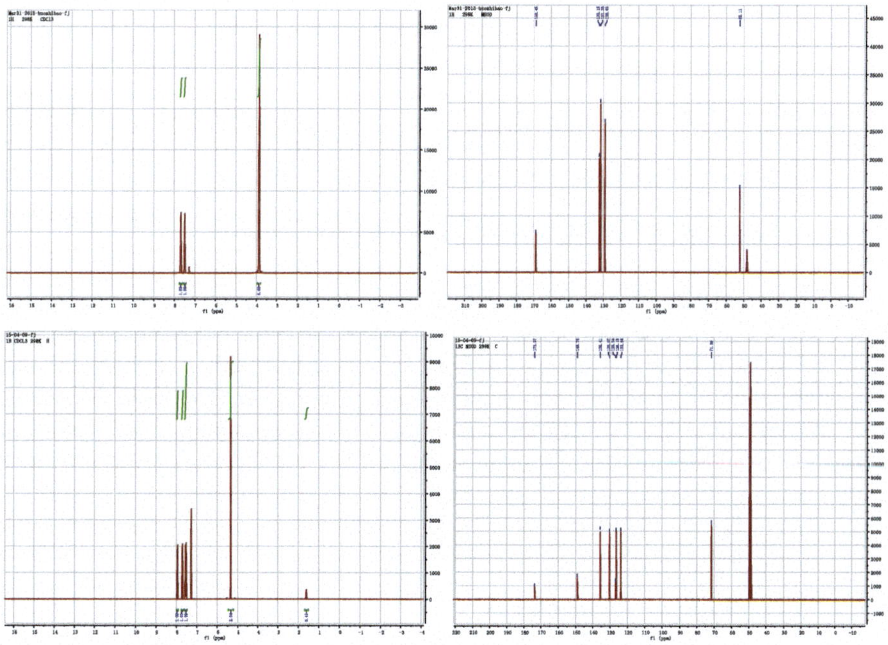

Fig. 9.4 ^{1}H NMR and ^{13}C NMR of DMP and PHT. Reprinted from Elsevier (Fu et al., 2019)

Table 9.1 Effect of catalysts on the yield of PHT

Entry	Reductant	Catalyst	Yield of PHT (%)
1	–	–	0
2	Zn	–	24
3	–	Cu	0
4	–	CuO	0
5	Zn	Fe	47
6	Zn	Cu	61
7	Zn	Co	43
8	Zn	CuO	85
9	Zn	CuO_2	55
10	Zn	Fe_2O_3	0
11	Zn	Fe_3O_4	26
12	Zn	TiO_2	57
13	Zn	MnO_2	58
14	Zn	ZrO_2	42
15	Zn	Cu Fe_2O_4	12
16	Zn	Pd/C	24
17	Zn	Ni	15

[a]Reprinted from Elsevier (Fu et al., 2019)

solid sample. Our previous work (Xu et al., 2014) had confirmed that CuO could be reduced completely into Cu by in situ-formed hydrogen by the oxidation of Zn in water. We measured the Cu ion concentration in the solution by ICP and only a small amount of Cu ions (< 1 ppm) in the liquid sample was found. This result suggested that Cu was most likely present in the solid residual (Fig. 9.6).

Next, we investigated the effect of the amount of CuO on the yield of PHT. The reaction was conducted in the presence of 25 mmol Zn with 25% water filling at 250 °C for 120 min. The results are shown in Fig. 9.7a. The yield of PHT increased significantly at first as the amount of CuO increased from 0 to 5 mmol, and the maximum value of 85% PHT was achieved at 5 mmol CuO. The yield decreased when the amount of CuO was further increased. These results may be led from that more CuO would lead to the decomposition of PHT to more by-products, which might result in a decrease in PHT selectivity.

9.3.2 *Effect of the Reductants*

Several active metals, such as Al, Mg, Fe, Mn and Zn, were investigated for a suitable reductant in the formation of PHT. Table 9.2 showed the effect of different active metals on the yields of PHT. Among the various active metals tested, it can be seen

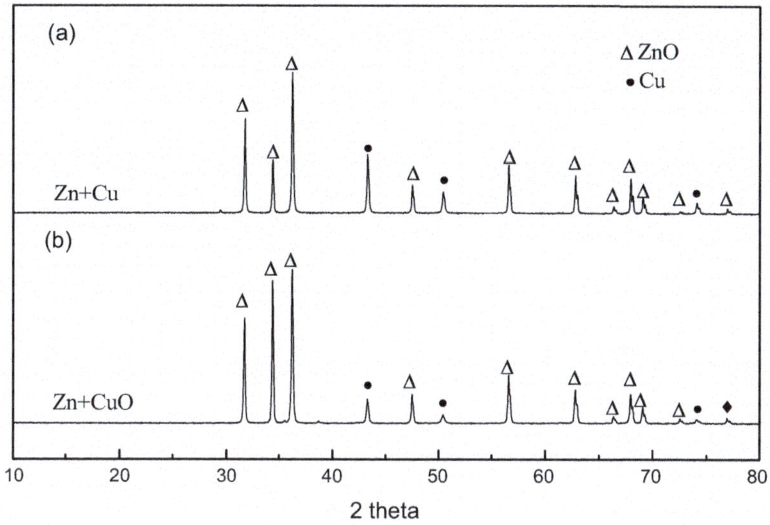

Fig. 9.5 XRD pattern of soil residual. (DMP: 0.5 mmol, Cu or CuO: 5 mmol, Zn: 25 mmol, water filling: 25%, 250 °C, 2 h). Reprinted from Elsevier (Fu et al., 2019)

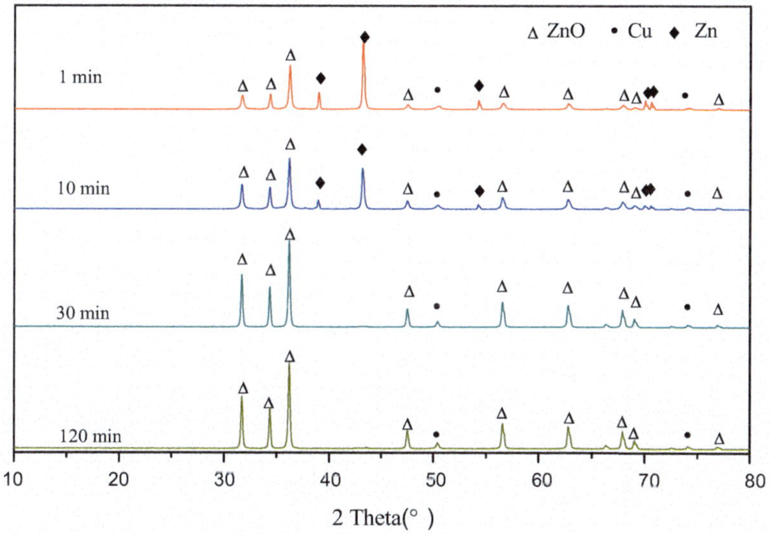

Fig. 9.6 XRD patterns of the solid samples. Reprinted from Elsevier (Fu et al., 2019)

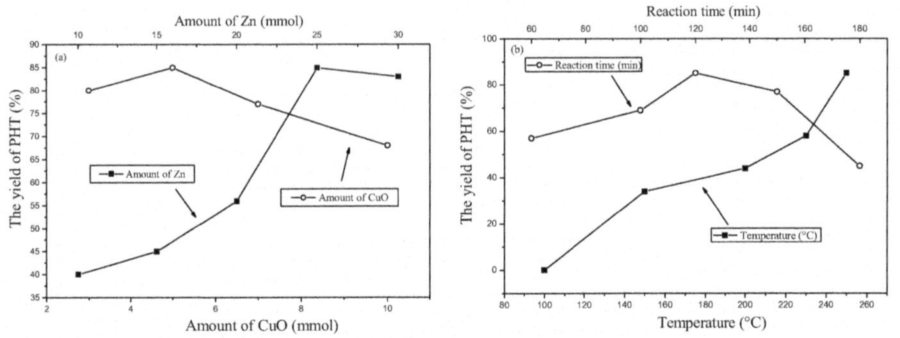

Fig. 9.7 **a** Effect of CuO and Zn amount. **b** Effect of temperature and time. Reprinted from Elsevier (Fu et al., 2019)

Table 9.2 Effect of reductants on the yield of PHT

Entry	Reductant	Yield of PHT (%)
1	–	0
2	Al	2
3	Mg	0.4
4	Fe	0.4
5	Zn	85
6	Mn	0.1

[a] Reprinted from Elsevier (Fu et al., 2019)

that Al, Mg, Fe and Mn were almost ineffective for the transformation, and gave a very lower yields (entries 2–4, 6), while Zn powder had a much higher selectivity for the conversion of DMP and gave the best 85% yield of PHT compared to Al, Mg, Mn, and Fe. Thus, Zn was chosen for the following experiments.

To monitor the oxidation process from Zn to ZnO during the reaction at different time, such as 1, 10, 30 and 120 min, the XRD patterns of the solid samples were shown in Fig. 9.1. Firstly, a quick oxidation from Zn to ZnO in water can be observed. After 1 min, peaks of Zn gradually become smaller until it disappeared at 30 min, whereas peaks of ZnO became bigger gradually as time increased. However, peak of Cu is also observed after the reaction. These results indicate that the DMP can be converted to PHT by in situ formed hydrogen by the oxidation of Zn in water over CuO.

Next, we investigated the effect of the amount of Zn on the yield of PHT. The reaction was conducted in the presence of 5 mmol CuO with 25% water filling at 250 °C for 120 min and with Zn as reductant, of which, the amounts is from 0 to 30 mmol, the results are shown in Fig. 9.5a. As we mentioned above, no PHT is observed without the addition of Zn. The yield of PHT increased significantly at first as the amount of Zn increased from 0 to 25 mmol, and the maximum value of 85% PHT is achieved at 25 mmol. The yield of PHT decreases somewhat when the amount of Zn exceeds 25 mmol. The results indicate that the improvement in hydrogen production with an increased amount of Zn in water is better for the conversion of DMP to PHT, but higher hydrogen partial pressure will cause the decomposition of PHT and decrease the yield.

9.3.3 *Effect of Other Parameters*

Figure 9.8 shows the effect of water filling from 15 to 45% on the yield of PHT with the conditions of 5 mmol CuO and 25 mmol Zn at 250 °C for 120 min. Due to the reason of safety, the max water filling is 45%. The yields of PHT are almost interestingly constant between the water filling of 15 and 20%. The yield remarkably increases from 20 to 25% and reaches the maximum value of 85%, and then decreases after 25%. Probably, decreasing yield is because a higher water filling causes the decomposition of PHT when water filling increases.

In the conversion of DMP, temperature was found to be a very influencing factor. The effect of temperature was investigated at different temperature values from 100 °C to 250 °C with the addition of 5 mmol CuO and 25 mmol Zn at water filling of 25% for 120 min, and the results are showed in Fig. 9.5b. The reaction did not give the desire PHT at 100 °C, indicating that a lower temperature was completely ineffective in the conversion of DMP. The yield of PHT increased continuously from

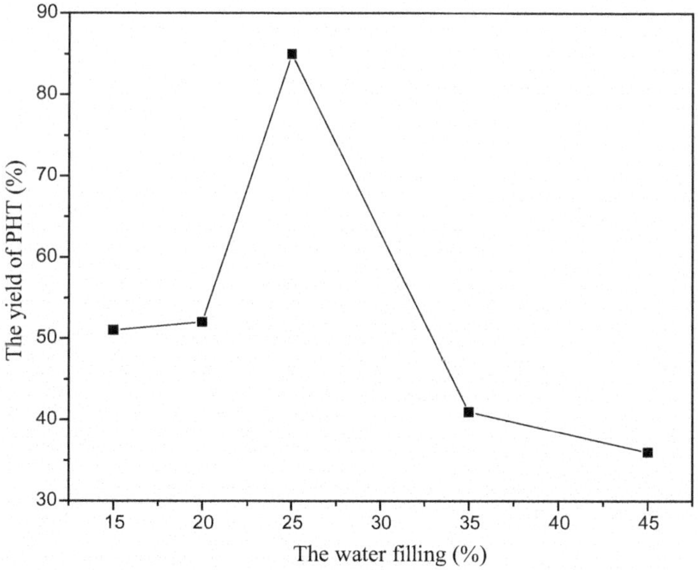

Fig. 9.8 Effect of water filling. Reprinted from Elsevier (Fu et al., 2019)

100 to 250 °C. Between 200 and 250 °C, the yields of PHT increased the most obviously. The maximum value of 85% PHT was obtained at 250 °C. However, we did not continue the optimization because 250 °C was the limited temperature of the Teflon material.

To examine the effect of different reaction time from 60 to 300 min on the yield of PHT, the results in Fig. 9.5b indicates that the reaction time plays a important role in the process. The yield of PHT increases at the first 120 min and the maximum value of 85% PHT is obtained. However, with further reaction time increases, the yield of PHT decreases. To investigate the reason of decreasing yield of PHT by using Cu and ZnO obtained after the reaction and PHT as a reactant, the reaction of prepared 50 mmol/L PHT was performed in the presence of 5 mmol Cu and 25 mmol ZnO with 25% water filling at 250 °C for 120 min. The GC-FID results indicate that the initial concentration of PHT has decreased from 50 mmol/L to 22 mmol/L. These results suggest that Cu and ZnO can cause the decomposition of PHT obtained at a long reaction time in the process (Larcher & Patrice, 2000).

Finally, a carbon balance was conducted at the optimal condition. The results show that the conversion of DMP for liquid products is approximately 98.2%. The main products observed for this transformation were PHT and methanol. In addition, CO_2 as main gaseous product with trace amounts of other gaseous products is also observed in the transformation (see Fig. 9.9).

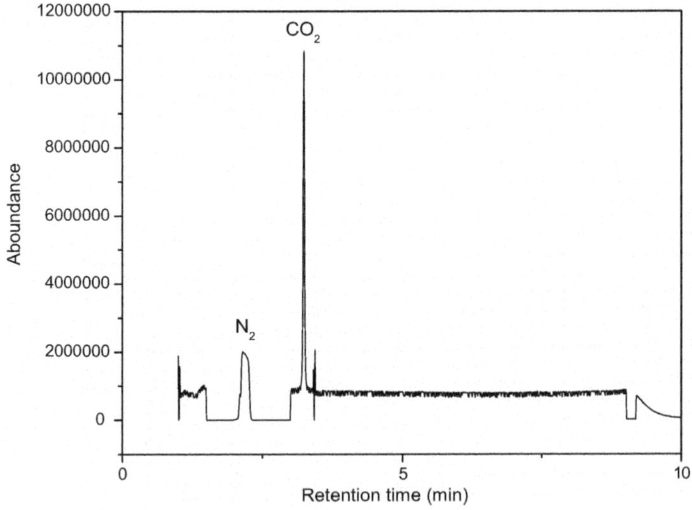

Fig. 9.9 GC-FID after the reaction. Reprinted from Elsevier (Fu et al., 2019)

9.3.4 *Investigation of Gases Hydrogen*

The effect of gases hydrogen on the formation of PHT from DMP in high temperature water was investigated. The reaction was conducted in the presence of 5 mmol CuO by using 2.8 MPa H_2 with 25% water filling at 250 °C for 2 h (Procedure in details, see Supporting Information). As a result, only peak of PHT was observed by GC-FID, and the PHT yield of 5.2% was much lower than the yield of 85% by in situ-formed hydrogen. The result suggested that gases hydrogen was not effective for the product of PHT from DMP in high temperature water.

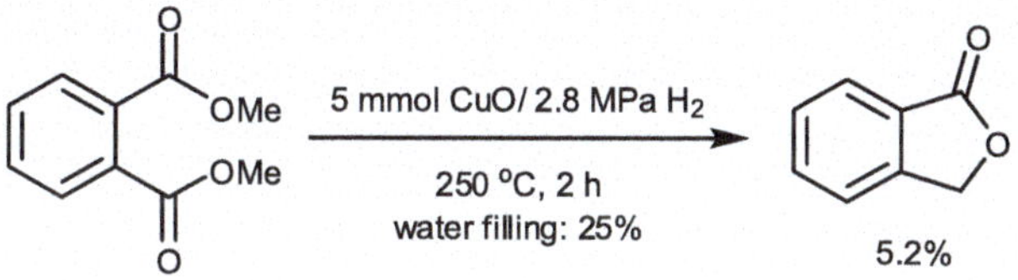

9.4 Mechanism Study

Initially, in order to investigate the pathway for the conversion of DMP, we examined the peak changes of intermediates in the process. At 250 °C we chose the 10 min, 30 min and 120 min to check the intermediates. As shown in Fig. 9.10. it is obvious that the peak of DMP is reducing with the increase of reaction time and gradually disappeares from 10 to 120 min, while desired PHT keeps increasing. However, we observed the intermediate 2-hydroxymethyl benzoic acid marked with circular on the right when the reaction time was 30 min though its peak was small.

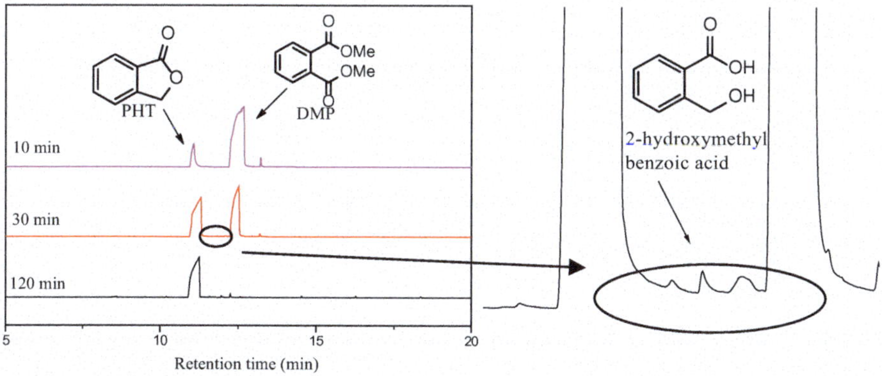

Fig. 9.10 GC–MS chromatogram of compounds during the reaction. Reprinted from Elsevier (Fu et al., 2019)

Based on previous reports and results in this study, the pathway for the conversion of DMP to PHT over CuO is proposed in Fig. 9.12 First, hydrogen is formed via the oxidation of Zn in water (Elliott, 2008). Then the in situ-formed hydrogen reacted with DMP molecule to obtain an intermediate 2-hydroxymethyl benzoic acid **5** (see Fig. 9.11) via intermediates **3** and **4** along with the loss of methanol (Kieboom & Rantwijk, 1977). Finally, the intramolecular cyclization of the intermediate 2-hydroxymethyl benzoic acid occurred and give the desired PHT.

Next, evidential experiment is conducted to investigate the feasibility of intermediate **5** in the process. We used the intermediate 2-hydroxymethyl benzoic acid **5** as the substrate to carry out the reaction at the optimized reaction conditions, and no surprisingly the expected PHT was produced in 17% yield (Fig. 9.13).

With the success in the conversion of DMP to PHT, we next theorized the conversion from dimethyl maleate (DMM) to butyrolactone could happen by this method. The reaction of DMM was carried out in the presence of 25 mmol Zn and 5 mmol

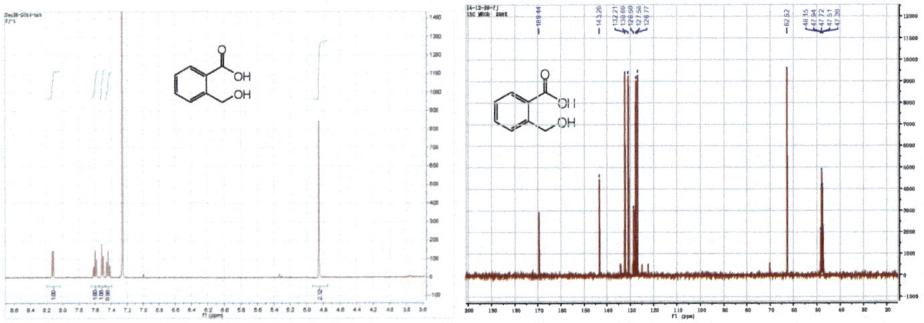

Fig. 9.11 ^{1}H NMR and ^{13}C NMR of DMP and PHT. Reprinted from Elsevier (Fu et al., 2019)

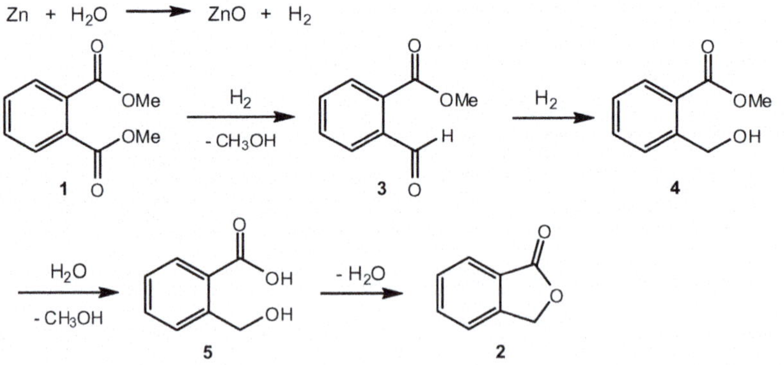

Fig. 9.12 Proposed pathway for the formation of PHT from DMP. Reprinted from Elsevier (Fu et al., 2019)

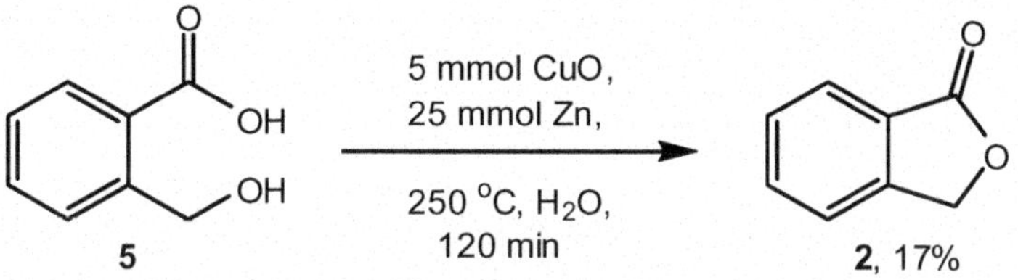

Fig. 9.13 Conversion of intermediate **5** to PHT. Reprinted from Elsevier (Fu et al., 2019)

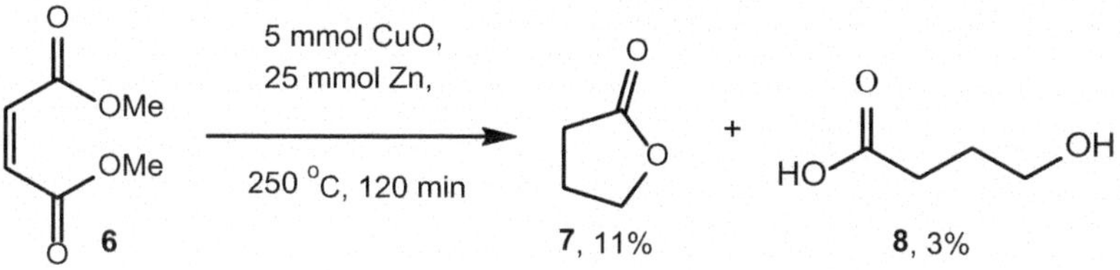

Fig. 9.14 Conversion of DMM to butyrolactone. Reprinted from Elsevier (Fu et al., 2019)

CuO with 25% water filling at 250 °C for 120 min and gave the expected butyrolactone **7** in 11% yield. The structure of butyrolactone was determined by the GC–MS spectrum as shown in Figs. 9.14 and 9.15.

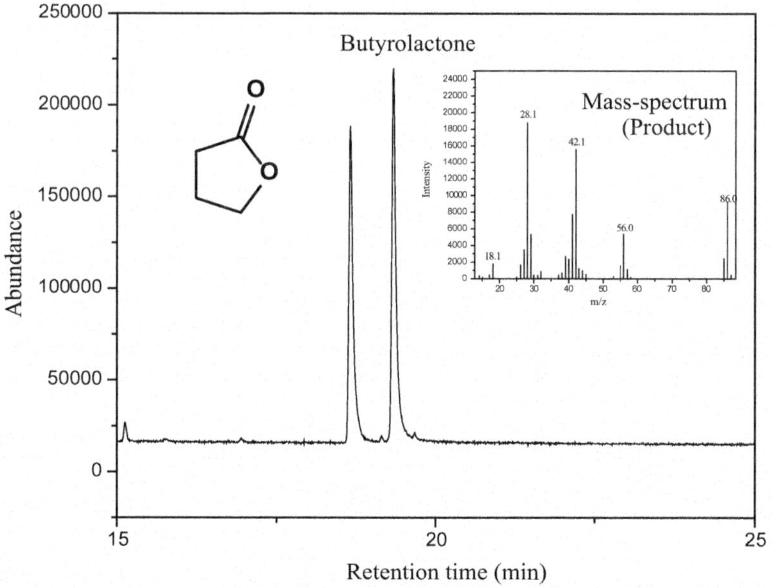

Fig. 9.15 GC–MS spectrum of butyrolactone. Reprinted from Elsevier (Fu et al., 2019)

9.5 Conclusions

We developed a highly efficient conversion of DMP to PHT over CuO in aqueous media. CuO catalyst exhibits high catalytic activity for the conversion of DMP. A higher yield of 85% was achieved in the presence of 25 mmol Zn and 5 mmol CuO with 25% water filling at 250 °C for 120 min. Mechanism study found that 2-hydroxymethyl benzoic acid as an intermediate was observed. Extended experiment in the conversion of DMM also produced butyrolactone at 11% yield in similar reaction conditions. Further study will be taken to develop efficient methods for the conversion of biomass to value-added chemicals.

References

Akiya, N., & Savage, P. E. (2002). Roles of water for chemical reactions in high-temperature water. *Chemical Reviews, 102*, 2725–2750.

Chheda, J. N., Huber, G. W., & Dumesic, J. A. (2007). Liquid-phase catalytic processing of biomass-derived oxygenated hydrocarbons to fuels and chemicals. *Angewandte Chemie International Edition, 46*, 7164–7183.

Clarke, M. L., Díaz-Valenzuela, M. B., & Slawin, A. M. Z. (2007). Hydrogenation of aldehydes, esters, imines, and ketones catalyzed by a ruthenium complex of a chiral tridentate ligand. *Organometallics, 26*, 16–19.

Corma, A., Iborra, S., & Velty, A. (2007). Chemical routes for the transformation of biomass into chemicals. *Chemical Reviews, 107*, 2411–2502.

Donner, C. D. (2013). Tandem Michael–Dieckmann/Claisen reaction of ortho-toluates—the Staunton-Weinreb Annulation. *Tetrahedron, 69*, 3747–3773.

Duan, P., & Savage, P. E. (2011). Upgrading of crude algal bio-oil in supercritical water. *Bioresource Technology, 102*, 1899–1906.

Elliott, D. C. (2008). Catalytic hydrothermal gasification of biomass. *Biofuels, Bioproducts and Biorefining, 2*, 254–265.

Fang, Z., Sato, T., Smith, R. L., Inomata, H., Arai, K., & Kozinski, J. A. (2008). Reaction chemistry and phase behavior of lignin in high-temperature and supercritical water. *Bioresource Technology, 99*, 3424–3430.

Fu, J., Ren, D., Li, L., Liu, Y., Jin, F., & Huo, Z. (2016). Efficient conversion of dimethyl phthalate to phthalide over CuO in aqueous media. *Catalysis Today, 263*, 123–127.

Hayashi, S., Asakawa, Y., Ishida, T., & Matsuura, T. (1967). Phthalate esters of Cryptotaenia Canadensis Dc. Var. Japonica Makino (Umbelliferae). *Tetrahedron Letters, 8*, 5061–5063.

He, C., Tian, G., Liu, Z., & Feng, S. (2010). A mild hydrothermal route to fix carbon dioxide to simple carboxylic acids. *Organic Letters, 12*, 649–651.

Hermann, B. G., Blok, K., & Patel, M. K. (2007). Producing bio-based bulk chemicals using industrial biotechnology saves energy and combats climate change. *Environmental Science and Technology, 41*, 7915–7921.

Hernández, E., Vélez, J. M., & Vlaar, C. P. (2007). Synthesis of 1,4-dihydro-benzo[D][1,3]oxazin-2-ones from phthalides via an aminolysis-Hofmann rearrangement protocol. *Tetrahedron Letters, 48*, 8972–8975.

Huber, G. W., & Corma, A. (2007). Synergies between bio- and oil refineries for the production of fuels from biomass. *Angewandte Chemie International Edition, 46*, 7184–7201.

Huber, G. W., Cortright, R. D., & Dumesic, J. A. (2004). Renewable alkanes by aqueous-phase reforming of biomass-derived oxygenates. *Angewandte Chemie, 116*, 1575–1577.

Huber, G. W., Chheda, J. N., Barrett, C. J., & Dumesic, J. A. (2005). Production of liquid alkanes by aqueous-phase processing of biomass-derived carbohydrates. *Science, 308*, 1446–1450.

Huber, G. W., Iborra, S., & Corma, A. (2006). Synthesis of transportation fuels from biomass: Chemistry, catalysts, and engineering. *Chemical Reviews, 106*, 4044–4098.

Huo, Z., Hu, M., Zeng, X., Yun, J., & Jin, F. (2012). Catalytic reduction of carbon dioxide into methanol over copper under hydrothermal conditions. *Catalysis Today, 194*, 25–29.

Jin, F., & Enomoto, H. (2011). Rapid and highly selective conversion of biomass into value-added products in hydrothermal conditions: Chemistry of acid/base-catalysed and oxidation reactions. *Energy & Environmental Science, 4*, 382–397.

Jin, F., Gao, Y., Jin, Y., Zhang, Y., Cao, J., Wei, Z., & Smith, R. L., Jr. (2011). High-yield reduction of carbon dioxide into formic acid by zero-valent metal/metal oxide redox cycles. *Energy & Environmental Science, 4*, 881–884.

Jin, F., & Heiji, E. (2009). Hydrothermal conversion of biomass into value-added products: Technology that mimics nature. *Bioresources, 4*.

Kieboom, A., & Rantwijk, F. V. (1977). Hydrogenation and hydrogenolysis in synthetic organic chemistry.

Kruse, A., & Dinjus, E. (2007). Hot compressed water as reaction medium and reactant: 2. Degradation reactions. *The Journal of Supercritical Fluids, 41*, 361–379.

Larcher, D., & Patrice, R. (2000). Preparation of metallic powders and alloys in polyol media: A thermodynamic approach. *Journal of Solid State Chemistry, 154*, 405–411.

Matteoli, U., Blanchi, M., Menchi, G., Prediani, P., & Piacenti, F. (1984). Homogeneous catalytic hydrogenation of dicarboxylic acid esters. *Journal of Molecular Catalysis, 22*, 353–362.

Mitchell, A. S., & Russell, R. A. (1995). Annulation reactions with stabilized phthalide anions. *Tetrahedron, 51*, 5207–5236.

Rathwell, K., & Brimble, M. A. (2007). Use of stabilized phthalide anion annulation reactions in synthesis: An update. *Synthesis, 2007*, 643–662.

Snieckus, V. (1990). Directed ortho metalation. Tertiary amide and O-carbamate directors in synthetic strategies for polysubstituted aromatics. *Chemical Reviews, 90*, 879–933.

Tian, G., Yuan, H., Mu, Y., He, C., & Feng, S. (2007). Hydrothermal reactions from sodium hydrogen carbonate to phenol. *Organic Letters, 9*, 2019–2021.

Wang, Y., Duan, X., Tian, D., Lu, G., & Yu, H. (2013). Throughput and delay limits of 802.11p and its influence on highway capacity. *Procedia—Social and Behavioral Sciences, 96*, 2096–2104.

Xiao, J., Huo, Z., Ren, D., Zhang, S., Luo, J., Yao, G., & Jin, F. (2015). A novel approach for 1,2-propylene glycol production from biomass-derived lactic acid. *Process Biochemistry, 50*, 793–798.

Xu, L., Huo, Z., Fu, J., & Jin, F. (2014). Highly efficient conversion of biomass-derived glycolide to ethylene glycol over CuO in water. *Chem. Comm., 50*, 6009–6012.

Yokoyama, T., & Fujita, N. (2004). Hydrogenation of dicarboxylic acid diesters to corresponding dialdehydes over ZRO_2 based catalysts (part 2) analysis of catalyst surface using diffuse reflectance FT-IR and CP/MAS 13c-NMR spectroscopy. *Journal of the Japan Petroleum Institute, 47*, 341–347.

Chapter 10
Heterogeneous Cu_2O-Mediated Ethylene Glycol Production from Dimethyl Oxalate

10.1 Introduction

Ethylene glycol (EG), as an important intermediate chemical and solvent, has often been found as antifreeze, lubricant, plasticizer, and surfactant, which own an important status in chemical industry (Xu et al., 1995; Yue et al., 2012).Due to its application value and the increasing industrial demand, the development of the efficient methods for the synthesis of EG is attracting great interest. Currently, the industrial synthetic methods of EG are mainly the oxidation of petroleum-derived ethylene (Chen et al., 2008; Othmer & Thakar, 1958). However, over-dependence on fossil fuels still has drawbacks. In view of problems, to develop a research towards more economical and efficient synthetic methods of EG is necessary.

In recent years, the hydrogenation of biomass-derived DMO to EG had been receiving increasing attention (Turek et al., 1994). However, DMO with a less electrophilic carbonyl group is a stable compound and the reaction to produce EG is still a great challenge (Fig. 10.1). Since the 1970s, some interesting results in the hydrogenation of DMO to EG had been reported by using ruthenium-based homogenous catalysts, such as $H_4Ru_4(CO)_8(PBu_3)_4$ (Matteoli et al., 1984, 1986) and $Ru(acac)_3$ (Van et al., 2003). In 2010, Dai et al. first reported a facile and efficient one pot route for the gas-phase selective hydrogenation of DMO to EG over the Ag/SiO_2 catalyst (Yin et al., 2009, 2010b). Recently, the supported copper-based catalysts, such as Cu/HMS (Yin et al., 2010a), CuB/HMS (Yin et al., 2011), Cu/SiO_2 (Miyazaki et al., 1986), $Cu–Cr_2O_3$ (Susumu et al., 1982) and $Cu/ZnO/Al_2O_3$ (Zhang et al., 2012), have been reported to exhibit a high catalytic activity for the hydrogenation of DMO to EG, and good conversion of DMO and selectivity to EG were obtained in the transformation. However, although these methods had proven to be extremely efficient in the hydrogenation of DMO to EG, most of them still have their numerous limitations. For instance, the use of high-pressure gaseous hydrogen, organic solvents, expensive prepared catalysts, and longer reaction times, etc. In addition, the use of Ru-based catalysts can form byproducts and separated difficultly products from catalysts, and the toxicity of Cr-containing Cu-based catalysts is harmful to humans and can cause a

Z. Huo, *Diverse Hydrogen Sources for Biomass-derivatives Conversion*,
https://doi.org/10.1007/978-981-99-1673-3_10

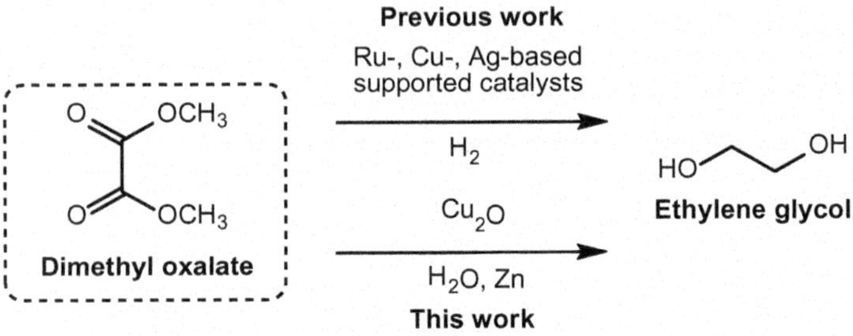

Fig. 10.1 Formation of EG from DMO. Reprinted from Elsevier (Li et al., 2016)

severe environmental pollution. Therefore, the development of an efficient synthetic process is highly desirable in the conversion of DMO to EG.

The development of economical hydrogen sources for the conversion of DMO to EG is essential. Previously, our studies (Huo et al., 2012; Jin & Heiji, 2009; Jin et al., 2011; Wang et al., 2015; Xiao et al., 2015; Xu et al., 2014) reported some interesting results in high temperature water. In this process, water acts not only as environmentally benign reaction medium but also as a source of hydrogen, which could be advantageous to avoid the use of gaseous hydrogen. For example, we reported an efficient method for the production of EG from glycolide in high yield over CuO in water (Xu et al., 2014). Furthermore, ZnO can be cycled to Zn through a ZnO/Zn redox process using solar energy (Robert & Service, 2009; Steinfeld, 2005; Stamatiou et al., 2013). Inspired by the findings above, we herein present a new strategy for the conversion of DMO to EG over heterogeneous Cu_2O in high temperature water (Fig. 10.1).

10.2 Experimental

10.2.1 Materials and Methods

10.2.1.1 Materials

Dimethyl oxalate (Aladdin Chemical, > 98%) was used as the initial reactant. EG (TCI Scientific Ltd, > 98%) and some compounds in the reactions such as ethanol and methanol (J&K Chemical, > 98%) were used as test materials. Zn power (AR) was obtained from Aladdin Chemical Reagent Co. Ltd. Cu_2O power (under 200-mesh in size) and other metals were obtained from Sinophram Chemical Reagent Co. Ltd, China.

10.2.1.2 Experimental Procedure

All reactions were performed in a Teflon reactor, whose internal volume is 30 mL covered by a metal reaction kettle. The experiment procedure is as follows. Firstly, the 7.5 mL water mixture of test materials with 0.5 mmol DMO, 25 mmol Zn, and 5 mmol CuO were loaded into the reactor. Then the reactor was put into a drying oven at a desired temperature. The reaction time was defined as the duration the reactor was kept in the oven. Water filling was defined as the ratio of the volume of the solution to the internal volume (30 mL) of the reactor. After a desired reaction time, the reactor was quickly moved out from the drying oven to cool down. Finally, after cooling off, the liquid samples were collected and filtered with 0.45 μm filter membrane for GC–MS and HPLC analysis, Solid samples were washed with deionized water and ethanol several times and dried in air for XRD analysis.

10.2.1.3 Analysis Methods

After the reactions, the qualitative analysis results of liquid samples were based on the GC–MS analysis (Agilent 7890A). And the quantitative estimation of EG and byproducts was based on average values obtained from the HPLC analysis (Agilent 1260 serials, VIS detector) of three samples. Details on the conditions of GC–MS/FID and HPLC analyses are available elsewhere (Jin et al., 2001). The EG yield is defined as the percentage of EG to initial DMO on the mole basis, calculated as following equation:

$$Y(\%) = \frac{mole\,of\,product}{initial\,mole\,of\,DMO} \times 100$$

10.3 Results and Discussion

Initially, a tentative experiment was conducted to investigate whether hydrogenation of DMO could produce EG in the presence of CuO by using 2.8 MPa H_2 with 25% water filling at 250 °C for 2 h. However, no desired EG was obtained, indicating that the use of gaseous hydrogen was not effective for the conversion of DMO in high temperature water. Thus, we focused our study on the use of in situ formed hydrogen for the synthesis of EG from DMO.

10.3.1 Metals or Metal Oxides Screening

The effect of metals or metal oxides was investigate whether DMO could be converted into EG in the presence of CuO and Zn with 25% water filling at 250 °C for 2 h. As a result, the HPLC chromatograms and GC–MS spectrum of the liquid samples showed that the desired EG appeared, and EG was the main product with little amounts of ethanol (Kong et al., 2015) and methanol as shown in Fig. 10.2a, b.

The effect of metals or metal oxides, such as Cu, Fe, Ni, CuO, and Cu_2O on the yield of EG was tested. The results showed in (Table 10.1) that the desired EG was obtained in 22%, 28%, 62% and 70% yields over Ni, Co, CuO, and Cu_2O, respectively (entries 6–9). Among them, Cu_2O exhibited a high catalytic performance for the conversion of DMO and the corresponding EG yield of 70% was obtained (entry 9). However, no desired EG was obtained when the other metals or metal oxides, such as Fe, Cu, Cr, Al_2O_3, and Fe_2O_3 were used (entries 4–5, 10–14). In addition, no desired EG was obtained in the absence of metals or metal oxides and/or reductant in the transformation (entries 1–3), it indicated that the importance of the combined use of both reductant and metals or metal oxides fairly crucial to this reaction.

To investigate the role of the metals or metal oxides in the conversion of DMO into EG, the XRD patterns of the solid residue after the reaction showed that Zn was oxidized to ZnO, and the Co and Ni still existed. This indicated that the Zn acted as a reductant and Co and Ni as catalysts (Fig. 10.3a, b). Unexpectedly, the Cu was observed in the solid residue instead of Cu_2O when Cu_2O was used, the most likely reason is that Cu_2O was reduced completely to Cu by in situ formed hydrogen via the oxidation of Zn in high-temperature water (Fig. 10.3a) (Zhang et al., 2012). To confirm this hypothesis, the reaction was conducted by using Zn and Cu_2O at the same reaction conditions without DMO. The XRD pattern after the reaction showed that ZnO and Cu were observed in solid residue. We investigated that if there are dissolved Cu present by ICP in liquid sample, and found only a small amount of Cu ions (<1 ppm). This result indicated that there was almost no Cu ion present in the process.

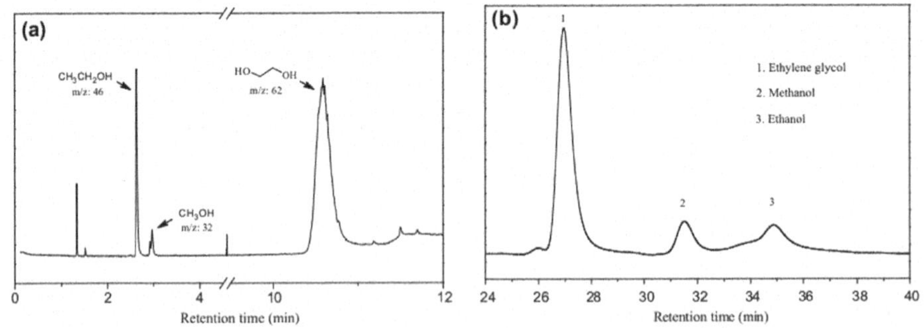

Fig. 10.2 **a** GC–MS spectrum of liquid samples after the reaction. **b** HPLC chromatograms of liquid samples after the reaction (Li et al., 2016)

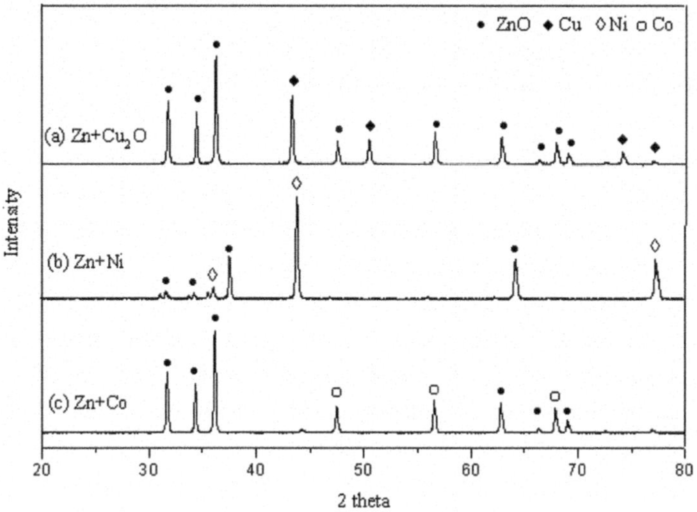

Fig. 10.3 **a**, **b**, **c** XRD patterns of the solid sample after the reaction. Reprinted from Elsevier (Li et al., 2016)

Table 10.1 Catalyst screening on the formation of EG[a]

Entry	Catalyst	Reductant	Yield (%)[b]	
			EG	Ethanol
1	–	–	0	0
2	CuO	–	0	0
3	–	Zn	0	0
4	Fe	Zn	0	0
5	Cu	Zn	0	0
6	Ni	Zn	22	18
7	Co	Zn	8	15
8	CuO	Zn	62	7
9	Cu_2O	Zn	70	6
10	Ni_2O_3	Zn	0	0
11	TiO_2	Zn	0	0
12	Fe_2O_3	Zn	0	0
13	Al_2O_3	Zn	0	0
14	SiO_2	Zn	0	0

Reprinted from Elsevier (Li et al., 2016)

[a] Reaction condition: DMO 0.5 mmol, catalyst 5 mmol, reductant 25 mmol, 250 °C, 2 h, water filling 25%

[b] The yield is calculated as the percentage of EG to the initial DMO on the mole number

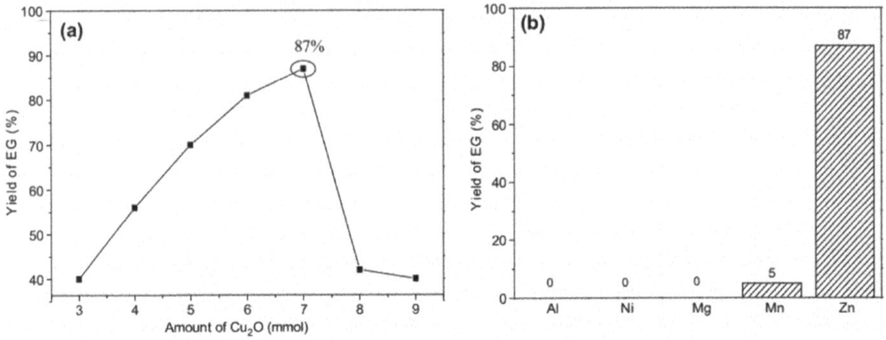

Fig. 10.4 **a** Effect of amount of Cu_2O (DMO 0.5 mmol, water filling 25%, 250 °C, 2 h. **b** Effect of reductants (DMO 0.5 mmol, Cu_2O 7 mmol, reductant 25 mmol, water filling 25%, 250 °C, 2 h (Li et al., 2016)

Next, we studied the effect of the amount of Cu_2O on the yield of EG by remaining other conditions unchanged. The result was shown in (Fig. 10.4a), the yield of EG increased quickly at first with the amount of Cu_2O increasing from 3 to 7 mmol, and the yield was decreased when the amount of Cu_2O exceeded 7 mmol. The decreasing yield might be that more Cu_2O would result in the decomposition of desired EG to ethanol or other compounds as fellows below (Larcher & Patrice, 2000). Hence, it can be confirmed that the best yield of 87% EG at 7 mmol Cu_2O was achieved.

$$CH_2OH{-}CH_2OH_{(l)} + 5Cu_2O_{(s)} \rightarrow 10Cu^0_{(s)} + 2CO_{2(g)} + 3H_2O_{(g)}$$

10.3.2 *Effect of Reductant*

To search for a suitable reductant for the formation of EG, several active metals were investigated. As can be seen in (Fig. 10.4b) and (Table 10.2), the use of Al, Ni and Mg were ineffective to this reaction, which no expected EG were obtained. When Mn was used as a reductant, it gave a low yield of EG of 6%. As a result, Zn powder exhibited the much higher reactivity for the conversion of DMO to EG compared to other metals, and the EG yield of 87% was obtained.

Then, the effect of the amount of Zn on the yield of EG was studied. The reaction was conducted at amount of Zn range from 0 to 35 mmol over 7 mmol Cu_2O with 25% water filling at 250 °C for 2 h. With the amount of Zn increased, the amount of hydrogen production by the oxidation of Zn increased as well. Figure 10.5a shows no EG was obtained when the amount of Zn increased from 0 to 10 mmol. When the amount of Zn increased from 15 to 30 mmol, the yield of EG increased quickly to a higher yield of 91% at 30 mmol Zn. However, with the amount of Zn was continuously increased, the yield of EG showed a decreasing trend, which may due

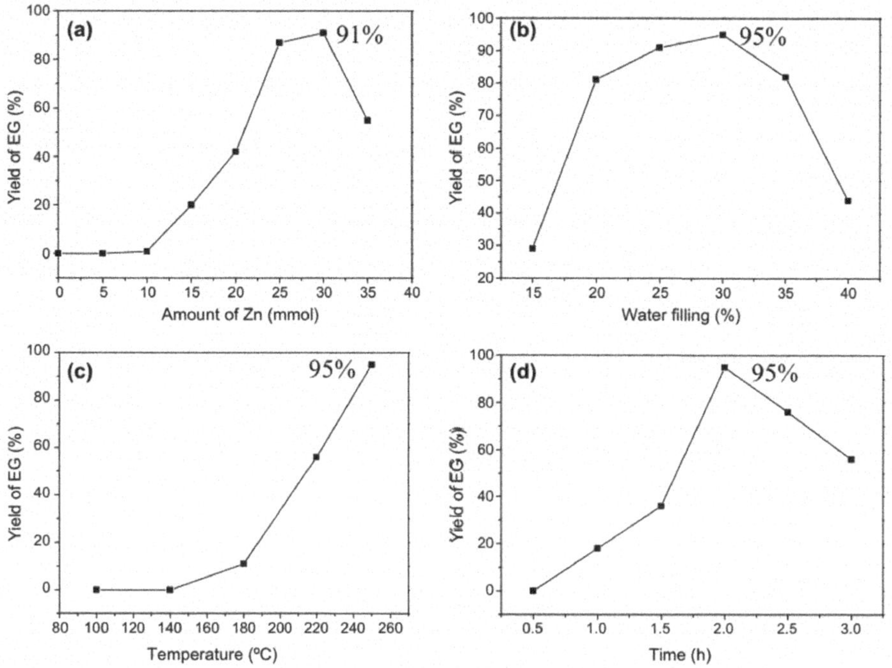

Fig. 10.5 Effect of several parameters on the formation of EG. **a** Amount of Zn; **b** Water filling; **c** Reaction temperature; **d** Reaction time. Reprinted from Elsevier (Li et al., 2016)

to the side reaction of EG with excess hydrogen. Thus, 30 mmol Zn was chosen as the suitable amount of reductant.

10.3.3 Effects of Water Filling, Reaction Temperature and Time

Effect of water filling on the yield of EG was investigated within the range from 15 to 40% with 7 mmol Cu_2O and 30 mmol Zn at 250 °C for 2 h. In (Fig. 10.5b), it can be see that EG yield increased quickly from 29 to 91% as the water filling increased from 15 to 25%. The yield of EG was improved to a maximum yield of 95% when water filling reached 30%. As water filling changed from 35 to 40%, the yield of EG decreased quickly. In this process, due to the maximum durable pressure limit of the Teflon reactor material, experiments at more than 40% water filling were not carried out.

As we know, temperature is one of the most important factors to affect a chemical reaction. In this study, experiments were carried out at a temperature range of 100 to 250 °C by using 7 mmol Cu_2O and 30 mmol Zn with water filling 30% for 2 h.

In (Fig. 10.5c), the yields of EG showed an interesting increase from 100 to 250 °C. Between 100 and 140 °C, no EG was observed. When temperature increased to180 °C, a lower yield of desired EG was obtained. The maximum EG value of 95% was achieved at the temperature of 250 °C. As the tolerated temperature of the Teflon reactor was 250 °C, the optimization at the temperature above 250 °C had not been continued.

Experiments were performed by changing the reaction time from 0.5 to 3 h in the presence of Cu_2O 7 mmol, Zn 30 mmol with 30% water filling at 250 °C to examine the effect of reaction time. The result in (Fig. 10.5d) showed that the EG yield increased as the reaction time increased from 0.5 to 2 h, and it reached the highest yield of 95% when the reaction time is 2 h. When the reaction time exceeded 2 h, the EG yield showed a downward trend. The decreasing yield of EG after 2 h might be attributed to the decomposition of produced EG. We carried out a reaction to confirm this assumption, and found that the concentration of 55.6 mmol/L EG was decreased to 50.4 mmol/L in the presence of 30 mmol ZnO and 7 mmol Cu with 30% water filling at 250 °C for 2 h.

10.3.4 Investigation of Active Species

As mentioned above, XRD pattern showed that Cu_2O was reduced completely to Cu by *in-situ* formed hydrogen via the oxidation of Zn in high-temperature water (Fig. 11.3a). Also, the results showed that the use of commercial Cu was completely ineffective for EG production (Table 10.1 entry5). To investigate the true catalytically active species Cu or Cu_2O, Cu isolated from a catalytic system after reaction (Cu_2O reduction), the experiment of DMO was performed in the presence of 7 mmol Cu and 30 mmol Zn with 30% water filling at 250 °C for 2 h. Interestingly, in situ formed Cu showed a high activity for DMO conversion to produce EG in 90% yield. Meanwhile, as shown in entry 3 in (Table 10.1), in situ formed ZnO did not show the catalytic performance for EG production. These results indicated that isolated Cu was a true catalytically active species (Cu_2O as a pre-catalyst).

Table 10.2 Effect of reductants on the formation of EG

Entry	Reductant	Yield (%)	
		EG	Ethanol
1	Al	0	0
2	Ni	0	0
3	Mg	0	0
4	Mn	5	0
5	Zn	87	6

Reaction conditions: DMO 0.5 mmol, Cu_2O 7 mmol, reductant 25 mmol, 250 °C, 2 h, water filling 25%. Reprinted from Elsevier (Li et al., 2016)

10.3.5 Recycle of Cu Formed in Situ

Next, the stability of in situ formed Cu was examined. In situ formed Cu after the reaction was reused as catalyst for DMO conversion. After four runs, the results were shown in (Fig. 10.6) and (Table 10.3), the yield of EG decreased from 95 to 78%, the yield of methanol decreased slightly while the yield of ethanol increased slightly. These results indicated that isolated Cu could still maintain the activity for the DMO conversion after four runs. Furthermore, SEM images of isolated Cu after the 1st and 3rd runs showed that no significant changes were observed (Fig. 10.7).

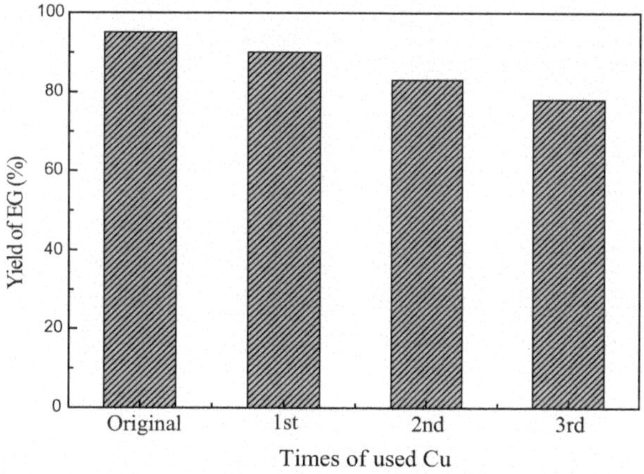

Fig. 10.6 Effect of reused Cu on the formation of EG. (DMO 0.5 mmol, isolated Cu 7 mmol, reductant 30 mmol, water filling 30%, 250 °C, 2 h). Reprinted from Elsevier (Li et al., 2016)

Table 10.3 Reaction conditions: 0.5 mmol DMO, 7 mmol isolated Cu, 30 mmol Zn, 30% water filling, 250 °C, 2 h

Entry	Reuse	Yield (%)		
		EG	MeOH	EtOH
1	Original	95	96	4
2	1	90	95	5
3	2	83	92	9
4	3	78	90	12

Reaction conditions: 0.5 mmol DMO, 7 mmol isolated Cu, 30 mmol Zn, 30% water filling, 250 °C, 2 h. Reprinted from Elsevier (Li et al., 2016)

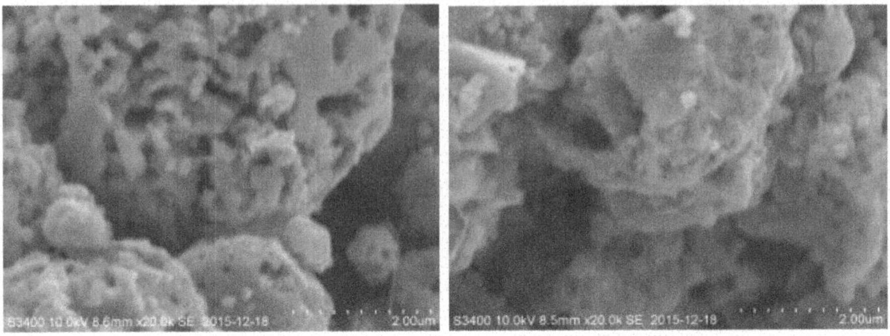

(a) 1st run (b) 3rd run

Fig. 10.7 SEM images of isolated Cu after the 1st and 3rd runs. Reprinted from Elsevier (Li et al., 2016)

10.3.6 Mass Balance

Finally, a carbon balance was investigated at the optimal condition, and found that the main liquid products were EG (95%) and methanol (96%) with a little amounts of ethanol (4%). The main gaseous products CO_2 and H_2 in the process were observed.

10.4 Proposed Mechanism

Previously, many researches had reported the proposed mechanism for EG synthesis via the hydrogenation of DMO by using gaseous hydrogen (Van et al., 2003). Based on the finding in previous and current studies, the pathway for the conversion of DMO to EG over Cu_2O was proposed in (Fig. 10.8). As mentioned above, in situ formed Cu was a true catalytically active species. First, the hydrogen was generated via the oxidation of Zn in high temperature water (Resende & Savage, 2010, Douglas et al., 2008). Then the in situ formed hydrogen was adsorbed on the surface of in situ formed Cu through the formation of hydrogen bonds between hydrogen and Cu. When the hydrogen reacted with DMO, an intermediate methyl glycolate was obtained, which subsequently reacted with in situ formed hydrogen on the surface of Cu to produce the desired EG. Finally, the ethanol was obtained through catalytic hydrogenolysis of produced EG.

To investigate the feasibility of the proposal above, an experiment of intermediate, methyl glycolate, was carried out at the optimized reaction condition, and the desired EG was obtained in 43% yield. Hence, this result suggested that methyl glycolate as an intermediate in the conversion of DMO into EG in high temperature water.

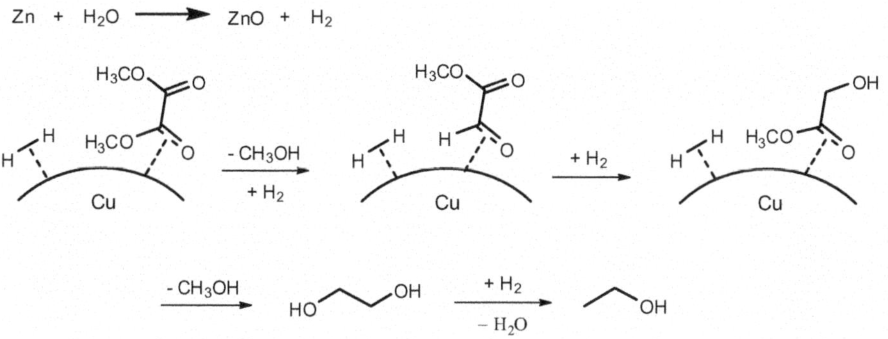

Fig. 10.8 Plausible pathway for the formation of EG. Reprinted from Elsevier (Li et al., 2016)

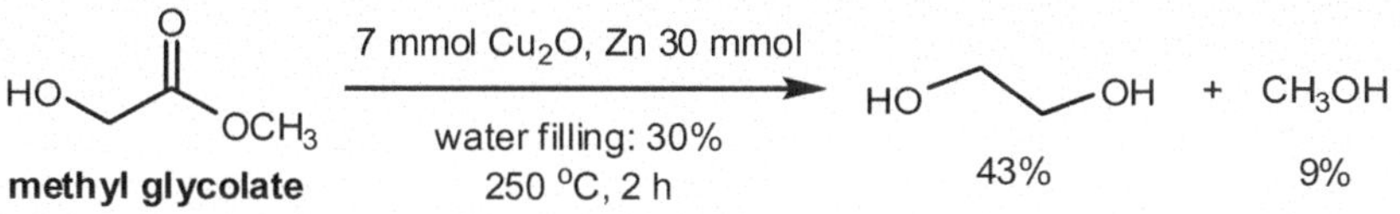

10.5 Conclusions

In this study, we developed an efficient route for the synthesis of EG from DMO over Cu_2O in high temperature water. By investigating various parameters, such as catalyst, reductant, temperature, time and water filling, etc. we obtained the optimal conditions for this transformation. In situ formed Cu as a true catalytically active species showed a good catalytic performance for the conversion of DMO to EG, and the maximum value of 95% was achieved in the presence of 30 mmol Zn and 7 mmol Cu_2O with 30% water filling at 250 °C for 2 h. Isolated Cu can be recycled four times with slight decreasing yields. In present process, DMO has been converted efficiently into EG, which provided a new thinking in synthesis of EG.

References

Chen, L. F., Guo, P. J., Qiao, M. H., Yan, S. R., Li, H. X., Shen, W., Xu, H. L., & Fan, K. N. (2008). Cu/Sio_2 catalysts prepared by the ammonia-evaporation method: Texture, structure, and catalytic performance in hydrogenation of dimethyl oxalate to ethylene glycol. *Journal of Catalysis, 257*, 172–180.

Douglas, C., & Elliott. (2008). Catalytic hydrothermal gasification of biomass. *Biofuel Bioproducts and Biorefining, 6*, 2694–2700.

Huo, Z., Hu, M., Xu, Z., Yun, J., & Jin, F. (2012). Catalytic reduction of carbon dioxide into methanol over copper under hydrothermal conditions. *Catalysis Today, 194*, 25–29.

Jin, F., & Heiji, E. (2009). Hydrothermal conversion of biomass into value- added products: Technology that mimics nature. *BioResources, 4*, 704.

Jin, F., Ying, G., Jin, Y., Zhang, Y., Cao, J., Zhen, W., & Smith Jr, R. (2011). High-Yield reduction of carbon dioxide into formic acid by zero-valent metal/metal oxide redox cycles. *Energy and Environmental Science, 4*, 881–884.

Jin, F. M., Kishita, A., Moriya, T., & Enomoto, H. (2001). Kinetics of oxidation of food wastes with H_2o_2 in supercritical water. *Journal of Supercritical Fluids, 19*, 251–262.

Kong, X., Zhang, X., & Chen, J. (2015). Highly active and selective Cocu/Zno catalysts prepared by mild oxalic acid co-precipitation method in dimethyl oxalate hydrogenation. *Catalysis Communications, 65*, 46–50.

Larcher, D., & Patrice, R. (2000). Preparation of metallic powders and alloys in Polyol media: A thermodynamic approach. *Journal of Solid State Chemistry, 154*, 405–411.

Li, L., Ren, D., Fu, J., Liu, Y., Jin, F., & Huo, Z. (2016). Heterogeneous Cu_2O-mediated ethylene glycol production from dimethyl oxalate. *Journal of Energy Chemistry, 25*, 507–511.

Matteoli, U., Bianchi, M., Frediani, P., Menchi, G., & Marchetti, M. (1984). Asymmetric synthesis by chiral ruthenium complexes. X. Enantioface discriminating isomerization of olefins. *Journal of Organometallic Chemistry, 263*, 243–246.

Matteoli, U., Menchi, G., Bianchi, M., & Piacenti, F. (1986). Homogeneous catalytic hydrogenation dicarboxylic acid esters II. *Journal of Organometallic Chemisty, 299*, 233–238.

Miyazaki, H., Uda, T., Hirai, K., Nakamura, Y., Ikezawa, H., & Tsuchie, T. (1986). Process for producing ethylene glycol and/or glycolic acid ester, catalyst composition used therefor, and process for production thereof. *Journal of Organometallic Chemistry, 299*, 233–238.

Othmer, D., & Thakar, M. (1958). Glycol production—Hydration of ethylene oxide. *Industrial and Engineering Chemistry Research, 50*, 1235–1244.

Resende, F., & Savage, P. E. (2010). Effect of metals on supercritical water gasification of cellulose and lignin. *Industrial and Engineering Chemistry Research, 49*, 2694–2700.

Robert, F., & Service, (2009). Solar Fuels. Biomass fuel starts to see the light.*Science, 326*, 1472–1475.

Stamatiou, A., Steinfeld, A., & Jovanovic, Z. R. (2013). On the effect of the presence of solid diluents during Zn oxidation By Co_2. *Industrial and Engineering Chemistry Research, 52*, 1859–1869.

Steinfeld, A. (2005). Solar thermochemical production of hydrogen—A review. *Solar Energy, 78*, 603–615.

Susumu, T., Kouzou, F., Keigo, N., Masaoki, M., & Katsuhiko, M. (1982). Continuous preparation of ethylene glycol.

Turek, T., Trimm, D. L., & Cant, N. W. (1994). The catalytic hydrogenolysis of esters to alcohols. *Catalysis Reviews: Science and Engineering, 36*, 645–683.

Van Engelen, M. C., Teunissen, H. T., & Elsevier, C. J. (2003). Suitable ligands for homogeneous ruthenium-catalyzed hydrogenolysis of esters. *Journal of Molecular Catalysis A-Chemical, 206*, 185–192.

Wang, J., Yao, G., Wang, Y., Zhang, H., Huo, Z., & Jin, F. (2015). A novel Pd/C-catalyzed conversion of glucose to 1,2-propanediol by water splitting with Zn. *RSC Advances, 5*, 51435–51439.

Xiao, J., Huo, Z., Ren, D., Song, Z., Jiang, L., Yao, G., & Jin, F. (2015). A novel approach for 1,2-propylene glycol production from biomass-derived lactic acid. *Process Biochemistry, 50*, 793–798.

Xu, G. H., Li, Y. C., Li, Z. H., & Wang, H. J. (1995). Kinetics of the hydrogenation of diethyl oxalate to ethylene glycol. *Industrial and Engineering Chemistry Research, 34*, 2371–2378.

Xu, L., Huo, Z., Fu, J., & Jin, F. (2014). Highly efficient conversion of biomass-derived glycolide to ethylene glycol over Cuo in water. *Chemical Communications, 45*, 6005–6012.

Yin, A., Guo, X., Dai, W.-L., & Fan, K. (2010a). Effect of Si/Al Ratio of mesoporous support on the structure evolution and catalytic performance of the Cu/Al-Hms catalyst. *Journal of Physical Chemistry, 114*, 8523–8532.

Yin, A., Guo, X., Dai, W., & Fan, K. (2010b). High activity and selectivity of Ag/Sio_2 catalyst for hydrogenation of dimethyl oxalate. *Chemical Communications, 46*, 4348–4350.

Yin, A., Qu, J., Guo, X., Dai, W. L., & Fan, K. (2011). The influence of B-doping on the catalytic performance of Cu/Hms catalyst for the hydrogenation of dimethyloxalate. *Applied Catalysis A-General, 400*, 39–47.

Yin, A. Y., Guo, X. Y., Dai, W. L., & Fan, K. N. (2009). The synthesis of propylene glycol and ethylene glycol from glycerol using Raney Ni as a versatile catalyst. *Green Chemistry, 11*, 1514–1516.

Yue, H., Zhao, Y., Ma, X., & Gong, J. (2012). Ethylene glycol: properties, synthesis, and applications. *Chemical Society Reviews, 41*, 4218–4244.

Zhang, S., Liu, Q., Fan, G., & Li, F. (2012). Highly-Dispersed copper-based catalysts from Cu–Zn–Al layered double hydroxide precursor for gas-phase hydrogenation of dimethyl oxalate to ethylene glycol. *Catalysis Letters, 142*, 1121–1127.

Chapter 11
Highly Efficient Conversion of Biomass-Derived Glycolide to Ethylene Glycol Over CuO in Water

11.1 Introduction

Ethylene glycol (EG) is an important intermediate chemical and solvent that has been widely used in antifreeze, liquid detergents, biodegradable polyester fibers, cosmetic products, pharmaceuticals, explosives, and plasticisers, among other products and often as a protecting group for carbonyl groups in organic synthesis. (Xu et al., 1995) Because of its industrial importance, the synthesis of ethylene glycol is of considerable interest. The current conventional process involves industrial production from petroleum-derived ethylene via hydration of the intermediate ethylene oxide. (Haas et al., 2005) However, over-consumption of fossil fuels and their limited supply has necessitated a reduction in our dependence on petroleum oil and the development of renewable and green energy sources. (Chheda et al., 2007; Hermann et al., 2007; Huber & Corma, 2007) Over the past several decades, biomass as a source for producing high value-added chemicals has attracted much attention due to its excellent properties: it is abundant, renewable, and produces less pollution, among other advantages. (Jin & Enomoto, 2011) The conversion of biomass-derived carbohydrates into 1,2-diols is currently receiving increasing attention. (Yin et al., 2009; Zhou et al., 2010).

The hydrogenation of esters to alcohols is an important reaction in organic chemistry. (Kuriyama et al., 2010)However, the reaction of ester groups, which have a less electrophilic carbonyl group, is still a great challenge of both academic and practical importance, especially for cyclic di-esters. (Ito et al., 2011) To date, very few studies have reported this transformation. Glycolides-important biomass-derived compounds, can be obtained from biomass sources such as sugar cane, sugarbeet, and immature grape juice. (Davachi et al., 2012) Recently, Milstein et al. reported the hydrogenation of biomass-derived glycolide to the corresponding ethylene glycol by well-defined electron rich bipyridine-based PNN [2-(di-tert-butylphosphinomethyl)-6-(diethylaminomethyl) pyridine] Ru (II) pincer complexes under mild conditions. (Balaraman et al., 2012) However, these reactions still have considerable drawbacks, including the use of high-pressure gaseous hydrogen (10–50 atm),

Z. Huo, *Diverse Hydrogen Sources for Biomass-derivatives Conversion*,
https://doi.org/10.1007/978-981-99-1673-3_11

which is not easy to obtain at a reduced energy cost; organic solvents (THF or 1,4-dioxane); expensive prepared catalysts (PNN-Ru); and longer reaction times (12–48 h). Therefore, the development of a suitable reaction process for the production of ethylene glycol from glycolide is highly desirable.

Previously, we and other groups reported a series of experiments investigating the conversion of CO_2 (He et al., 2010; Huo et al., 2012; Jin et al., 2011; Tian et al., 2007) and biomass (Chheda et al., 2007; Huber et al., 2004, 2005) into value-added chemicals in high-temperature water. In this process, water acts not only as an excellent environmentally benign reaction medium but also as a source of hydrogen generated by the reduction of cheap metal reductants. (Kruse & Dinjus, 2007) Recently, some research groups have also reported that ZnO can be cycled to Zn through a ZnO/Zn redox process using solar energy. (Service, 2009; Steinfeld, 2005; Stamatiou et al., 2013) On the basis of these findings, we reasoned that glycolide might produce ethylene glycol in a similar manner. The results obtained showed that the hypothesis proved feasible, and the reaction of glycolide proceeded effectively in the presence of CuO to give the desired ethylene glycol in high yield with high selectivity (Eq. 11.1). The detailed results are reported herein.

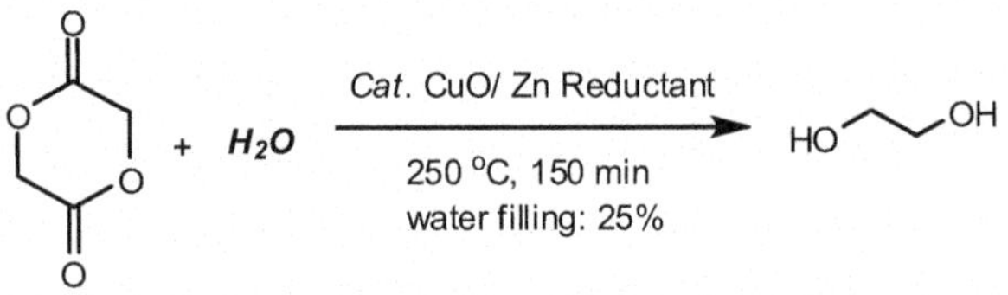

(11.1)

11.2 Experimental

11.2.1 General Information

Glycolide (≥ 99%, Sigma-Aldrich) was used as reagent and ethylene glycol (≥ 99.5%, GC) was purchased from TCI for the qualitative analysis of the products in the liquid samples. As preliminary tests, various active metals including Zn, Mg, Fe, Al, Mn (AR, Sinopharm Chemical Reagent Co., Ltd) were used as reductants in powder form, metal oxide including Fe_3O_4, Fe_2O_3, Ni_2O_3, TiO_2, ZrO_2, Cu_2O and CuO (150–200 mesh, i.e. 0.074–0.1 mm) (AR, Sinopharm Chemical Reagent Co., Ltd) were used as catalysts in powder form.

11.2.2 Product Analysis

After reaction, the liquid samples were analyzed with a GC-FID (Shimadzu GC-2010) equipped with an DB-FFAP capillary column with dimensions of 30 m × 250 μm × 0.25 μm for the quantification of ethylene glycol. The liquid samples' analyses were confirmed by GC–MS (Aglient GC7890A-MS5975C) equipped with an HP Innowax polyethylene glycol capillary column with dimensions of 30 m × 250 μm × 0.25 μm. Thin layer chromatography (TLC) was performed on aluminum-precoated plates of silica gel 60 with an HSGF254 indicator and visualized under UV light or developed by immersion in the solution of 0.6% $KMnO_4$ and 6% K_2CO_3 in water. The solid samples were characterized by X-ray diffraction (XRD) (Shimadzu XRD-6100) to determine the composition and phase purity.

The yield of ethylene glycol is defined as the carbon mole ratio of produced ethylene glycol to the initial glycolide as follows below. The yields were obtained from experiments over three times and relative error was less than 5%.

$$\text{Yield, mmol \%} = \frac{\text{C in ethylene glycols, mmol}}{\text{C in the initial glycolide, mmol}} \times 100(\%)$$

11.2.3 General Procedure for the Synthesis of Ethylene Glycol from Glycolide

Experiments were conducted in a Teflon-lined stainless steel batch reactor with an inner volume of 30 mL. The typical procedure for the synthesis of ethylene glycol from glycolide was as follows. First, glycolide (0.058 g, 0.5 mmol), Zn (1.63 g, 25 mmol), CuO (0.48 g, 6 mmol) and ultrapure water (7.5 mL) were loaded into reactor. Next, nitrogen was charged into reactor in order to exclude the effect of air and then the reactor was sealed and placed into an oven that had been preheated to the desired temperature. After 150 min, the reactor was removed from the oven and cooled to room temperature with an electric fan. The reaction time was defined as the time when the reactor's temperature was up to 250 °C. Water filling was defined as the ratio of the volume of the ultrapure water (7.5 mL) put into the reactor to the inner volume (30 mL) of the reactor. Final, after cooling off, liquid sample was collected and filtered with 0.45 μm Syringe Filter, the highest yield of ethylene glycol (94%) was obtained. Solid sample was collected and washed with deionized water and ethanol several times to remove impurities and dried in the oven at 50 °C for 24 h.

11.2.4 General Procedure for the Synthesis of 1,2-Propanediol from DL-Lactide

Experiments were conducted in a Teflon-lined stainless steel batch reactor with an inner volume of 30 mL. The typical procedure for the synthesis of ethylene glycol from glycolide was as follows. First, DL-lactide (0.072 g, 0.5 mmol), Zn (1.63 g, 25 mmol), CuO (0.48 g, 6 mmol) and ultrapure water (7.5 mL) were loaded into reactor. Next, nitrogen was charged into reactor in order to exclude the effect of air and then the reactor was sealed and placed into an oven that had been preheated to the desired temperature. After 150 min, the reactor was removed from the oven and cooled to room temperature with an electric fan. The reaction time was defined as the time when the reactor's temperature was up to 250 °C. Water filling was defined as the ratio of the volume of the ultrapure water (7.5 mL) put into the reactor to the inner volume (30 mL) of the reactor. Final, after cooling off, liquid sample was collected and filtered with 0.45 μm Syringe Filter, the highest yield of 1,2-propanediol (84%) was obtained. Solid sample was collected and washed with deionized water and ethanol several times to remove impurities and dried in the oven at 50 °C for 24 h.

11.3 Results and Discussion

Initially, we carried out an experiment to investigate the effectiveness of metal or metal oxides in the synthesis of ethylene glycol from glycolide in the presence of 6 mmol Cu and 20 mmol Zn at 250 °C for 150 min in water (water filling: 25%). The reaction proceeded efficiently and produced the desired ethylene glycol at a 23% yield, as determined by GC-FID and GC–MS (entry 4). The solid residual remaining after the reaction was analysed by XRD (Fig. 11.1), which showed that Cu was still present in the solid samples, while Zn was oxidised to ZnO. In this case, Zn acts as a reductant, and Cu acts as a catalyst for the conversion of glycolide. Next, we screened various catalysts to determine the most suitable one. The results are summarised in Table 11.1. The reaction did not give the desired ethylene glycol at all in the absence of both reductant and catalyst or in the presence of the catalyst only (entries 1 and 3). The reaction in the presence of 20 mmol Zn, without the addition of any catalyst, gave the desired ethylene glycol at a 9% yield (entry 2), indicating the importance of the combined use of both reductant and catalyst. Among the various catalysts we investigated (entries 5–13), Cu_2O and CuO improved the transformation efficiency and afforded ethylene glycol at a good yield with no glycolide recovery observed by thin layer chromatography (TLC) (entries 12 and 13); CuO gave much better results. A low yield was observed with the catalysts Fe_2O_3, Fe_3O_4, and Ni_2O_3 (entries 7, 8 and 9), which no reaction occurred when the catalysts Fe, Co, TiO_2, or ZrO_2 were used (entries 5, 6, 10 and 11).

The residual solid sample remaining after reaction in the presence of CuO and Zn was analysed by XRD. It can be seen that Cu and ZnO existed in solid samples in

Table 11.1 Effect of catalysts on the yields of ethylene glycol[a]

Entry	Reductant	Catalyst	Yield(%)[b]
1	–	–	0
2	Zn	–	9
3	–	Cu	0
4	Zn	Cu	23
5	Zn	Fe	0
6	Zn	Co	0
7	Zn	Fe_2O_3	14
8	Zn	Fe_3O_4	10
9	Zn	Ni_2O_3	10
10	Zn	TiO_2	0
11	Zn	ZrO_2	0
12	Zn	CuO	45
13	Zn	Cu_2O	37

Reprinted from ChemComm (Xu et al., 2014)
[a] Reaction conditions: glycolide: 0.5 mmol; Zn: 20 mmol; catalyst: 6 mmol; water filling: 25%; time: 150 min temp: 250 °C. [b] The yield is calculated as the percentage of ethylene glycol to the initial glycolide on the carbon base

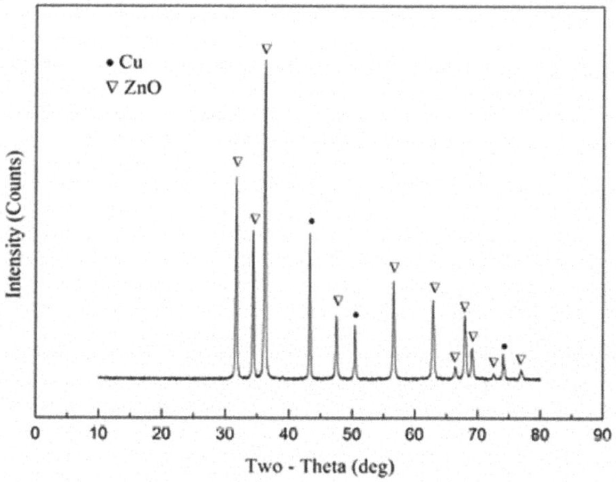

Fig. 11.1 XRD patterns of solid samples after reaction (glycolide: 0.5 mmol; Zn: 20 mmol; Cu (or CuO): 6 mmol; water filling: 25%; time: 150 min; temp: 250 °C). Reprinted from ChemComm (Xu et al., 2014)

Fig. 11.1. Unexpectedly, the Cu was observed in the solid sample instead of CuO. From this result, we reasoned that CuO as a catalyst might be reduced completely to Cu by in situ-formed hydrogen via the oxidation of Zn in high-temperature water. (Li et al., 2001) We measured the Cu ion concentration in the solution by ICP and found only a small amount of Cu ions (<1 ppm) in the liquid sample. This result suggested that Cu was most likely present in the solid residual. Thus, we performed the experiment under the same reaction conditions without the addition of glycolide to investigate the role of in situ-formed hydrogen. The results suggested that our hypothesis was completely realistic. CuO was reduced completely to Cu by in situ-formed hydrogen, and Cu and ZnO were found in the solid samples by XRD analysis after reaction.

Next, we investigated the effect of the amount of CuO on the yield of ethylene glycol. The reaction was conducted in the presence of 20 mmol Zn with 25% water filling at 250 °C for 150 min and 3 to 12 mmol CuO, the results are shown in Fig. 11.2. The yield of ethylene glycol increased significantly at first as the amount of CuO increased from 0 to 6 mmol, and the best yield of 45% ethylene glycol was achieved at 6 mmol CuO. The yield decreased as the amount of CuO was further increased. This may be because more CuO would lead to more cracking products such as acetic acid, CO_2 and H_2O (Eq. 11.2), which might result in a decrease in glycolide selectivity. (Larcher & Patrice, 2000)

$$CH_2OH - CH_2OH_{(l)} \rightarrow 5Cu^0_{(s)} + 2CO_2 + 3H_2O_{(g)} \tag{11.2}$$

Finally, a carbon balance was conducted. The results showed that the conversion of glycolide for liquid products was approximately 98%. The main product observed for this transformation was ethylene glycol, with a yield of 45%, other products included acetic acid, which had a 5% yield, and unknown compounds, CO_2 was the gas product observed in the transformation.

Experiments were performed using the various active metals, such as Zn, Al, Fe, Mn, and Mg, to search for a suitable reductant for the conversion of glycolide to ethylene glycol. Figure 11.3a shows the effect of different reductants on the yields of ethylene glycol. Among the various metals tested, Zn was highly effective in the conversion of glycolide and gave the best yield of ethylene glycol (45%) compared to Al, Mg, Mn, and Fe. Thus, Zn was chosen for the following experiments.

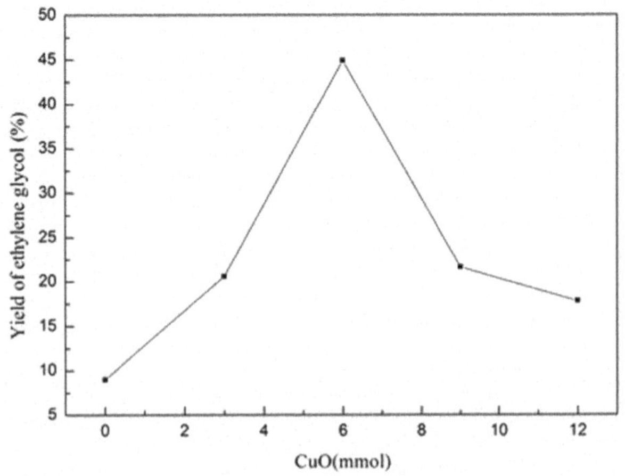

Fig. 11.2 Effect of the amount of CuO on the yields of ethylene glycol (glycolide: 0.5 mmol; Zn: 20 mmol; temp: 250 °C; time: 150 min; water filling: 25%). Reprinted from ChemComm (Xu et al., 2014)

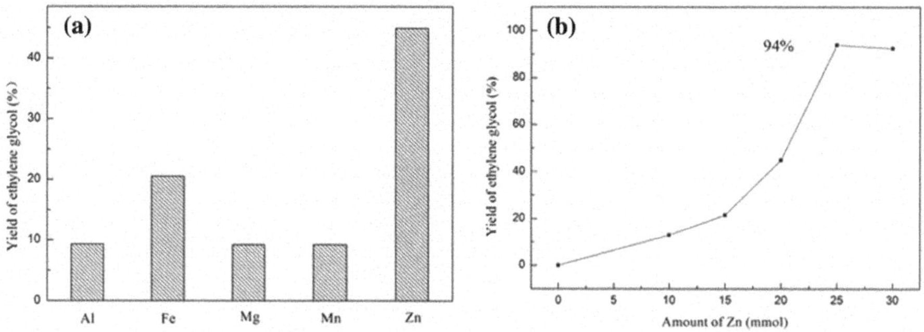

Fig. 11.3 **a** Effect of reductants (glycolide: 0.5 mmol; Metal: 20 mmol; CuO: 6 mmol; water filling: 25%; time: 150 min; temp: 250 °C). **b** Effect of the amount of Zn (glycolide: 0.5 mmol; CuO: 6 mmol; time: 150 min; temp: 250 °C; water filling: 25%). Reprinted from ChemComm (Xu et al., 2014)

Figure 11.3b shows the effect of the amount of Zn on the yield of ethylene glycol, using concentrations from 0 to 30 mmol. No ethylene glycol was observed without the addition of Zn. The yields of ethylene glycol improved gradually as the amount of Zn was increased from 0 to 15 mmol, then remarkably increased from 15 to 25 mmol. The yield of ethylene glycol decreased somewhat when the amount of Zn exceeded 25 mmol. The results indicated that the improvement in hydrogen production with an increased amount of Zn in high temperature water is in favourable for the conversion of glycolide to ethylene glycol. The optimal amount of Zn was 25 mmol, yielding ethylene glycol at a maximum value of 94%.

The effect of water filling on the yield of ethylene glycol from glycolide was investigated at different water filling values from 15 to 45% with the addition of 6 mmol CuO and 25 mmol Zn at 250 °C for 150 min. The results are shown in Fig. 11.4. First, the ethylene glycol yield increased slightly when the water filling was changed from 15 to 20%. The yield of ethylene glycol significantly increased as the water filling increased from 15 to 25% and then decreased after 25%. The optimum water filling was 25% obtaining the highest yield of 94% ethylene glycol from glycolide. The decrease in conversion of glycolide at a higher water filling is most likely due to a decreased concentration of the initial starting materials when water filling increased.

The effect of different temperatures from 100 to 250 °C on the yields of ethylene glycol was examined. The results are shown in Fig. 11.5a. The reaction did not take place at 100 °C, and no ethylene glycol was observed. This finding indicated that a lower reaction temperature was completely ineffective in the conversion of glycolide. The yield of ethylene glycol improved gradually as the reaction temperature was increased gradually from 100 °C to 200 °C, and then remarkably increased as the temperature increased from 200 to 250 °C. The maximum value of 94% ethylene glycol was obtained at 250 °C. However, the temperature was limited to 250 °C due to the maximum tolerated temperature of the Teflon material.

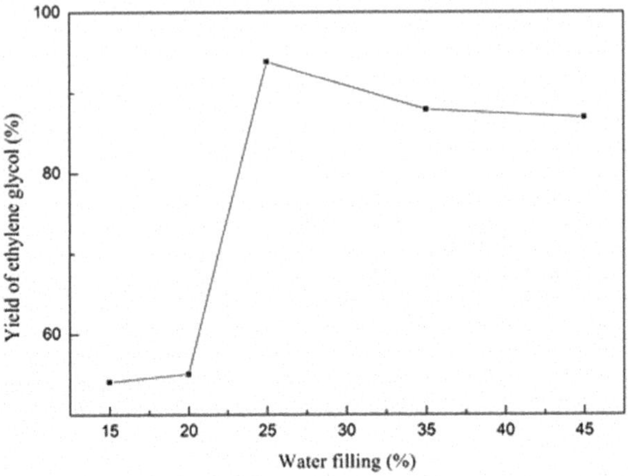

Fig. 11.4 Effect of water filling on the yields of ethylene glycol (glycolide: 0.5 mmol; Zn: 25 mmol; CuO: 6 mmol; time: 150 min; temp: 250 °C). Reprinted from ChemComm (Xu et al., 2014)

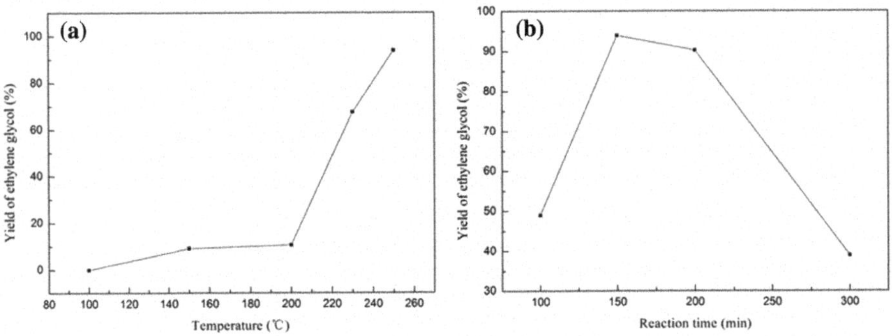

Fig. 11.5 Effect of reaction temperature and reaction time on the yields of ethylene glycol (**a** glycolide: 0.5 mmol; Zn: 25 mmol; CuO: 6 mmol; time: 150 min; water filling: 25%; **b** glycolide: 0.5 mmol; Zn: 25 mmol; CuO: 6 mmol; temp: 250 °C; water filling: 25%). Reprinted from ChemComm (Xu et al., 2014)

Experiments were performed by changing the reaction time from 100 to 300 min at 250 °C to examine the effect of reaction time on the yield of ethylene glycol. Figure 11.5b indicates that the yield increased drastically with increasing reaction time for the first 150 min, the maximum yield was obtained at 150 min, with a value of 94%. However, the yield of ethylene glycol decreased as the time was further increased. The decreasing yield of ethylene glycol after 150 min can be attributed to the decomposition of produced ethylene glycol. To confirm this assumption, the reaction of ethylene glycol was performed in the presence of 6 mmol Cu and 25 mmol ZnO with 25% water filling at 250 °C for 300 min. The GC-FID results indicated that the initial concentration of ethylene glycol 120 mmol/L was decomposed and decreased to 52 mmol/L, XRD analysis also showed that Cu and ZnO had not changed and were still present in the solid residual. These results suggest that a long reaction time has an influence on the yield of ethylene glycol obtained in the reaction in the presence of Cu and ZnO. (Larcher & Patrice, 2000).

Based on previous reports (Davachi et al., 2012) and the findings in this work, the pathway for the conversion of glycolide to ethylene glycol over CuO in high temperature water (HTW) was proposed in Fig. 11.6. First, hydrogen was formed via the oxidation of Zn in high temperature water. (Elliott, 2008; Resende & Savage, 2010) The in situ-formed hydrogen was adsorbed on the surface of Cu/CuO though the formation of hydrogen bonds between hydrogen and Cu/CuO. Additionally, glycolide was adsorbed on the surface of Cu/CuO by the combination of Cu/CuO with carbonylic C and O atoms. (Kieboom & Rantwiik, 1981; Yan et al., 2009) When the hydrogen reacted with glycolide molecules, an intermediate 2-hydroxyacetaldehyde was obtained. Next, both the intermediate 2-hydroxyacetaldehyde and the hydrogen formed were adsorbed on the surface of Cu/CuO, first, H was added to C = 11.

O and an intermediate linked by α-bond was formed, then the intermediate acquired one more H from the surface of Cu/CuO and the desired ethylene glycol was obtained.

Following the success in developing the conversion of glycolide to ethylene glycol in high temperature water, we next examined the scope of the reaction with respect to DL-lactide, and the results are shown in Eq. 11.3. The reaction of DL-lactide was carried out in the presence of 25 mmol Zn and 6 mmol Cu with 25% water filling at 250 °C for 150 min and gave the desired 1,2-propanediol at an 84% yield.

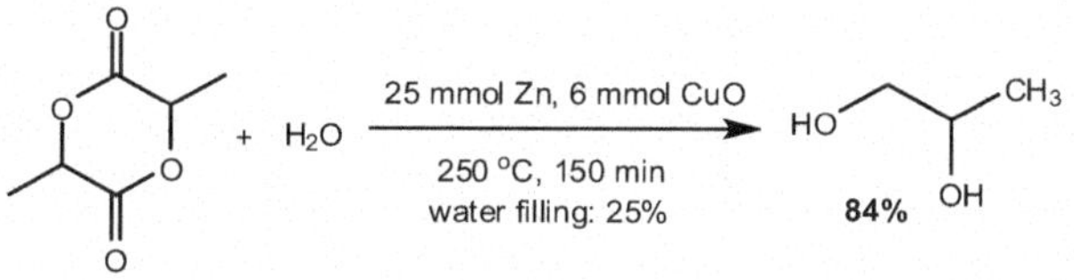

(11.3)

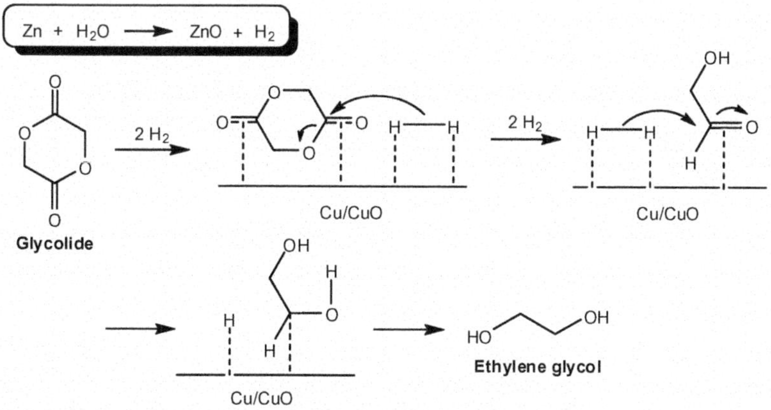

Fig. 11.6 Proposed mechanism for the formation of ethylene glycol

11.4 Conclusions

In summary, we first developed a highly efficient conversion for the synthesis of ethylene glycol from biomass-derived glycolide over CuO in high temperature water. With optimized conditions in hand, the maximum value of 94% ethylene glycol was achieved in the presence of 25 mmol Zn and 6 mmol CuO with 25% water filling at 250 °C for 150 min. The conversion of DL-lactide also produced 1,2-propanediol at 84% yield under optimal reaction conditions. This study provided a significant process for the conversion of biomass-derived cyclic di-esters to 1,2-diols with high selectivity and high yield. Further research is underway to develop efficient methods for the conversion of biomass-derived compounds to value-added chemicals in water.

References

Balaraman, E., Fogler, E., & Milstein, D. (2012). efficient hydrogenation of biomass-derived cyclic di-esters to 1,2-Diols. *Chemical Communications, 48*, 1111–1113.

Chheda, J. N., Huber, G. W., & Dumesic, J. A. (2007). Liquid-Phase catalytic processing of biomass-derived oxygenated hydrocarbons to fuels and chemicals. *Angewandte Chemie, 46*, 7164–7183.

Davachi, S. M., Kaffashi, B., & Roushandeh, J. M. (2012). Synthesis and characterization of a novel terpolymer based on llactide, glycolide, and trimethylene carbonate for specific medical applications. *Polymers for Advanced Technologies, 23*, 565–573.

Elliott, D. C. (2008). Catalytic hydrothermal gasification of biomass. *Biofuels, Bioproducts Biorefining, 2*, 254–265.

Haas, T., Jaeger, B., Weber, R., Mitchell, S. F., & King, C. F. (2005). New Diol Processes: 1,3-Propanediol and 1,4-Butanediol. *Applied Catalysis, A: General, 280*, 83–88.

He, C., Tian, G., Liu, Z., & Feng, S. (2010). A mild hydrothermal route to fix carbon dioxide to simple carboxylic acids. *Organic Letters, 12*, 649–651.

Hermann, B. G., Blok, K., & Patel, M. K. (2007). Producing bio-based bulk chemicals using industrial biotechnology saves energy and combats climate change. *Environmental Science and Technology, 41*, 7915–7921.

Huber, G. W., Chheda, J. N., Barrett, C. J., & Dumesic, J. A. (2005). Production of liquid alkanes by aqueous-phase processing of biomass-derived carbohydrates. *Science, 308*, 1446–1450.

Huber, G. W., & Corma, A. (2007). Synergies between bio- and oil refineries for the production of fuels from biomass. *Angewandte Chemie International Edition, 46*, 7184–7201.

Huber, G. W., Cortright, R. D., & Dumesic, J. A. (2004). Renewable alkanes by aqueous-phase reforming of biomass-derived oxygenates. *Angewandte Chemie International Edition, 43*, 1549–1551.

Huo, Z., Hu, M., Zeng, X., Yun, J., & Jin, F. (2012). Catalytic reduction of carbon dioxide into methanol over copper under hydrothermal conditions. *Catalysis Today, 194*, 25–29.

Ito, M., Ootsuka, T., Watari, R., Shiibashi, A., Himizu, A., & Ikariya, T. (2011). Catalytic hydrogenation of carboxamides and esters by well-defined Cp*Ru complexes bearing a protic amine ligand. *Journal of the American Chemical Society, 133*, 4240–4242.

Jin, F., & Enomoto, H. (2011). Rapid and highly selective conversion of biomass into value-added products in hydrothermal conditions: chemistry of acid/base-catalysed and oxidation reactions. *Energy & Environmental Science, 4*, 382–397.

Jin, F., Gao, Y., Jin, Y., Zhang, Y., Cao, J., Wei, Z., & Smith Jr, R. L. (2011). High-Yield reduction of carbon dioxide into formic acid by zero-valent metal/metal oxide redox cycles. *Energy & Environmental Science, 4*, 881–884.

Kieboom, A. P. G. & Rantwiik, F. V. (1981). Beijing, China.

Kruse, A., & Dinjus, E. (2007). Hot compressed water as reaction medium and reactant. *Journal of Supercritical Fluids, 41*, 361–379.

Kuriyama, W., Ino, Y., Ogata, O., Sayo, N., & Saito, T. (2010). A homogeneous catalyst for reduction of optically active esters to the corresponding chiral alcohols without loss of optical purities. *Advanced Synthesis & Catalysis, 352*, 92–96.

Larcher, D., & Patrice, R. (2000). Preparation of metallic powders and alloys in polyol media: A thermodynamic approach. *Journal of Solid State Chemistry, 154*, 405–411.

Li, Y., Xu, G.-H., Liu, C.-J., Eliasson, B., & Xue, B.-Z. (2001). Co-Generation of syngas and higher hydrocarbons from Co_2 and Ch_4 using dielectric-barrier discharge: effect of electrode materials. *Energ. Fuel., 15*, 299–302.

Resende, F. L. P., & Savage, P. E. (2010). Effect of metals on supercritical water gasification of cellulose and lignin. *Industrial and Engineering Chemistry Research, 49*, 2694–2700.

Service, R. F. (2009). Solar Fuels. Sunlight In Your Tank.*Science,* 326, 1472-1475

Stamatiou, A., Steinfeld, A., & Jovanovic, Z. R. (2013). On the effect of the presence of solid diluents during Zn oxidation by Co2. *Industrial and Engineering Chemistry Research, 52*, 1859–1869.

Steinfeld, A. (2005). Solar thermochemical production of Hydrogen—A review. *Solar Energy, 78*, 603–615.

Tian, G., Yuan, H., Mu, Y., He, C., & Feng, S. (2007). Hydrothermal reactions from sodium hydrogen carbonate to phenol. *Organic Letters, 9*, 2019–2021.

Xu, G.-H., Li, Y.-C., Li, Z.-H., & Wang, H.-J. (1995). Kinetics of the hydrogenation of diethyl oxalate to ethylene glycol. *Industrial and Engineering Chemistry Research, 34*, 2371–2378.

Xu, L., Huo, Z., Fu, J., & Jin, F. (2014). Highly efficient conversion of biomass-derived glycolide to ethylene glycol over CuO in water. *Chemical Communications, 50*(45), 6009–6012.

Yan, Z.-P., Lin, L., & Liu, S. (2009). Synthesis of Γ-Valerolactone by hydrogenation of biomass-derived levulinic acid over Ru/C catalyst. *Energ. Fuel., 23*, 3853–3858.

Yin, A. -Y., Guo, X. -Y., Dai, W. -L. & Fan, K. -N. (2009). The synthesis of propylene glycol and ethylene glycol from glycerol using raney Ni as a versatile catalyst. *Green Chemistry,, 11.*

Zhou, J., Guo, L., Guo, X., Mao, J. & Zhang, S. (2010). Selective hydrogenolysis of glycerol to propanediols on supported Cu-containing bimetallic catalysts. *Green Chemistry, 12.*

Part IV
Sodium Borohydride as Hydrogen Source

Chapter 12
A Supported Ni Catalyst Produced from Ni–Al Hydrotalcite-Like Precursor for Reduction of Furfuryl Alcohol to Tetrahydrofurfuryl Alcohol by $NaBH_4$ in Water

12.1 Introduction

Tetrahydrofurfuryl alcohol (THFA) have been widely used in various chemical industrial processes, including solvents (Suzuki et al., 2011), synthetic raw materials (Sato et al., 2013; Lee et al., 2013; Koso et al., 2009), antifreeze etc. (Lower et al., 1995). In general, most of industrial chemicals derived from fossil energy resources. However, with the depletion of fossil fuels, the conversion of renewable biomass to value-added chemicals has become more urgent (Corma et al., 2007; Huber et al., 2006). The current promising strategy to synthesize THFA is to hydrogenate furfuryl alcohol (FA), which is a important product obtained via hydrogenation of furfural. Note that furfural can be produced on a large scale over a simple acid catalysis process of lignocellulosic biomass (Karinen et al., 2011; Zhao et al., 2020; Wang et al., 2019). As such, increased interest has focused on this area. However, present methods for the conversion of FA into THFA remains a challenging issue, such as the use of flammable and explosive hydrogen (Sang et al., 2016; Khan et al., 2011; Wei et al., 2004), organic solvents (DMC (Yuan et al., 2015; King et al., 2015), THF (Geilen et al., 2011), pentanol (Jiao et al., 2014) etc.), noble metal catalysts such as (Cui et al., 2016; Panagiotopoulou & Vlachos, 2014; Gowda et al., 2012; Tike & Mahajani, 2007; Zhang et al., 2012), (Yuan et al., 2015; Biradar et al., 2014; Gong et al., 2014) and Pt (Wei et al., 2004), whilst longer reaction time (Geilen et al., 2011; Cui et al., 2016) and higher reaction temperature (Hronec et al., 2014; Chen et al., 2007; Yang et al., 2013; Hu et al., 2015). Therefore, development of new synthetic methods or efficient catalysts for the conversion of FA into THFA are still high desirable.

Layered double hydroxides (LDHs) are made up of positively charged brucite-like layers and charge compensating anions in interlayer region (Wang et al., 2012; Chen et al., 2017). Layered double oxides (LDOs) derived from LDHs possess the following advantages: high surface area, high dispersion, high loading, stable structure and high reactive activity (Fan et al., 2014). Therefore, it is widely used in biomass pyrolysis (Gao et al., 2017), pollutant disposal (Lei et al., 2017; Ma et al., 2016; Ju et al., 2017), adsorption (Wang et al., 2016, Peng et al., 2016), catalytic

Z. Huo, *Diverse Hydrogen Sources for Biomass-derivatives Conversion*,
https://doi.org/10.1007/978-981-99-1673-3_12

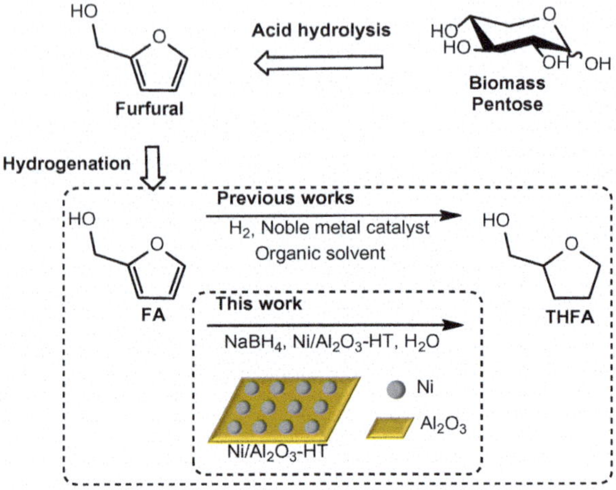

Fig. 12.1 Reduction of FA to THFA. Reprinted from Wiley (Ren et al., 2021)

oxidation (Zhang et al., 2015; Lin et al., 2017), catalytic hydrogenation (Cao et al., 2016) etc. Recently, Ni-based LDHs and LDOs catalysts have been reported for catalytic hydrogenation reaction and displayed excellent performances (Zhang et al., 2019; Zhang et al., 2020; Cao et al., 2019; Dong et al., 2019). At present, high pressure and explosive gaseous hydrogen is main reductant for reducing FA into THFA. Therefore, the use of safe reductant to replace gaseous hydrogen is much desirable. $NaBH_4$, a strong and green reducing agent, can reduce metallic ions (Fang et al., 2014; Furusawa et al., 2016), aldehydes (Ghaderi & Setamdideh, 2014), ketones (Zhou et al., 2004), and esters (Das et al., 2004; Khanapure et al., 2000). Unlike gaseous hydrogen, the transportation and storage of $NaBH_4$ is more convenient and safety due to the active ingredient of $NaBH_4$ is hydride ion (H–). To the best of our knowledge, $NaBH_4$ has not been reported to use as reductant to reduce FAHerein, we report a Ni-based LDOs (referred as Ni/Al_2O_3–HT) derived from Ni–Al hydrotalcite-like compound (referred as Ni–Al–HT) as an efficient heterogeneous catalyst for the reduction of FA using $NaBH_4$ as hydrogen donor at ambient conditions. Under the optimal conditions, the combination of Ni/Al_2O_3–HT and $NaBH_4$ exhibits good performance not only in reducing furan ring, but also in reducing ketones and benzene ring (Fig. 12.1).

12.2 Results and Discussion

Catalyst screening: Initially, a series of experiments were carried out to search for the suitable catalyst for the production of THFA, and the results were summarized in Table 12.1. The yields of THFA were determined by GC–FID. At first, transition metals (Fe, Cu, Mn, Co, Ni) were introduced as active centers to prepare metal/Al_2O_3-HT (mole ratio between metal and Al was 3:1). The results revealed that Ni/Al_2O_3–HT gave the highest THFA yield of 92.7% (entry 1), while Fe/Al_2O_3-HT, Cu/Al_2O_3–HT, Mn/Al_2O_3–HT and Co/Al_2O_3–HT were not effective or only afforded trace amount desired products (entries 2–5).In order to investigate the impact of catalyst supports, Ni based CeO_2, Bi_2O_3, ZrO_2 and Cr_2O_3 were prepared with same synthetic method of Ni/Al_2O_3-HT and displayed the yields of THFA were 4.2%, 28.6%, 33.6% and 56.5%, respectively (entries 6–9). According to soft and hard acids and bases rule (Pearson et al., 1963), we noticed that the yields and the hardness of metal ions had the same trend. In other words, the yield would be high if the metal ion of support was hard acid, and the yield would be low if the metal ion of support was soft acid. This might be due to the hard acid had stronger ability to decrease the electron density of furan ring, which enhanced the interaction between hydride ion and furan ring. This will be discussed in mechanism section later. Specific surface area is also an important factor for catalytic activity. In general, catalysts with higher specific area expose more active sites, which explains the highly activity of Ni/Al_2O_3-HT possessing the high specific surface area (145.6 m^2/g) (Table 12.2). Other catalysts such as Ni powder, Ni–Al alloy and Ni–Fe alloy were also investigated. However, only trace amount of products were obtained (entries 10–12, Table 12.1). The use of Raney Ni gave the yield of 83.9%, which was lower than Ni/Al_2O_3-HT (entry 13). In addition, Ni/Al_2O_3 prepared with co-precipitation and impregnation methods referred as Ni/Al_2O_3–CP and Ni/Al_2O_3–IMP, were also investigated, but only gave 70.4% and 57.3% yields of THFA, respectively (entries 14–15).

X-ray diffraction (XRD) patterns of Ni/Al_2O_3 prepared by different methods were shown in Fig. 12.2a. Metallic Ni and Ni oxide can be observed in Ni/Al_2O_3–HT, c. In order to understand the surface chemical environment of catalysts, X-ray photoelectron spectroscopy (XPS) was conducted. As shown in Fig. 12.3, the results of survey spectra display that the content of Al is much higher than Ni in all three catalysts indicating that Ni and NiO particles may imbed in the Al_2O_3. In addition, the amount of Ni on the surface of Ni/Al_2O_3–IMP is low demonstrating commercial γ–Al_2O_3 cannot absorb enough Ni ion via impregnation method to provide larghe amount of active sites. Concerning the BET data of three catalyst in Table 12.2 (entries 5, 7, 8), only hydrotalcite-like precursor derived Ni/Al_2O_3–HT could offer plenty of active sites, and Ni/Al_2O_3–CP and Ni/Al_2O_3–IMP either with low specific surface area, or the loading of active metal is low. The high-resolution Ni $2p_{3/2}$ XPS spectra of Ni/Al_2O_3–HT, Ni/Al_2O_3-CP and Ni/Al_2O_3–IMP were deconvoluted into three peaks, which can be assigned to Ni^0, Ni^{2+} and Ni satellite peaks (Fig. 12.2b). It illustrates that species of Ni on the surface of Ni/Al_2O_3 are metallic and oxidation states.

Transmission electron microscopy (TEM) and high-resolution TEM tests were also conducted. As shown in Fig. 12.4, obvious particle structure can be observed on Ni/Al_2O_3 prepared via hydrothermal and co-precipitation methods (Fig. 12.4a, c), and these particles are surrounded by amorphous materials which implies Ni and NiO imbed into Al_2O_3, which was consistent with the results of XPS. The morphology of

Table 12.1 Catalyst screening

Entry	Catalyst	THFA Yield (%)
1	Ni/Al_2O_3–HT	92.7
2	Co/Al_2O_3–HT	12.0
3	Fe/Al_2O_3–HT	Trace
4	Cu/Al_2O_3–HT	Trace
5	Mn/Al_2O_3–HT	Trace
6	Ni/CeO_2–HT	56.5
7	Ni/Bi_2O_3–HT	4.2
8	Ni/ZrO_2–HT	28.6
9	Ni/Cr_2O_3–HT	33.6
10	Ni powder	Trace
11	Ni–Al alloy	Trace
12	Ni–Fe alloy	3.0
13	Raney Ni	83.9
14	Ni/Al_2O_3–CP	70.4
15	Ni/Al_2O_3–IMP	57.3

Reaction conditions: 20 μL FA, 10 mg catalyst, 15 mg $NaBH_4$, 100 °C, 2 mL H_2O, 90 min. Reprinted from Wiley (Ren et al., 2021).

Table 12.2 Physicochemical properties of the catalysts

Entry	Catalyst	BET (m^2/g)	Pore diamter (nm)	Pore volume (cc/g)
1	Ni/Bi_2O_3–HT	6.5	1.9095	0.021
2	Ni/ZrO_2–HT	7.7	1.9102	0.031
3	Ni/Cr_2O_3–HT	41.8	3.8927	0.146
4	Ni/CeO_2–HT	18.9	1.9101	0.153
5	Ni/Al_2O_3–HT	145.6	8.7008	0.716
6	Raney Ni	34.2	1.7065	0.078
7	Ni/Al_2O_3–CP	39.9	1.7095	0.098
8	Ni/Al_2O_3–IMP	146.5	2.8096	0.214

BET surface area determined from N_2 gas adsorption. Reprinted from Wiley (Ren et al., 2021).

Ni/Al_2O_3–IMP in Fig. 12.4e, f were nanorods and sheets, however, barely particles were founded which means the low loading of Ni.

Effects of Ni/Al ratio, reduction temperature and solvents: Because Ni/Al_2O_3–HT exhibited excellent performance, the influence of Ni/Al ratio was examined. At first, the change of Ni/Al ratio didn't affect the specific surface areas of Ni/Al_2O_3–HT greatly (Table 12.3). As shown in Table 12.4, the highest yield of 100% was obtained when the ratio was 2:1 with 10 mg Ni/Al_2O_3–HT at reduction temperature 500 °C (entry 4). Higher or lower ratio of Ni/Al decrease the yield of THFA. The yield

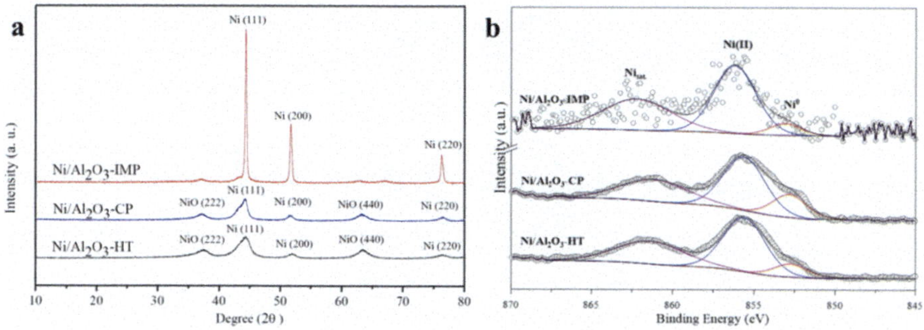

Fig. 12.2 **a** XRD patterns of Ni/Al_2O_3–HT, Ni/Al_2O_3–CP and Ni/Al_2O_3–IMP. 13.2 **b** Ni $2p_{3/2}$ XPS spectra of Ni/Al_2O_3–HT, Ni/Al_2O_3–CP and Ni/Al_2O_3–IMP. Reprinted from Wiley (Ren et al., 2021)

Fig. 12.3 Survey XPS spectra of Ni/Al_2O_3–HT,Ni/Al_2O_3–CP and Ni/Al_2O_3–IMP. Reprinted from Wiley (Ren et al., 2021)

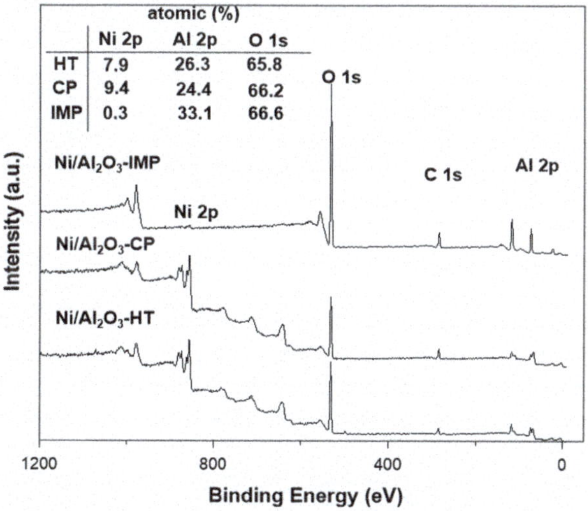

decreased with the decreasing amount of catalyst (entry 6). Therefore, the Ni/Al ratio of 2:1 and 10 mg catalyst were selected as reaction condition in follow-up experimentsIn Table 12.4, further experiments were carried out to optimize the best calcination temperature (entries 5,7–10). As expected, too low temperature such as 300 °C would lead to incomplete reduction of Ni oxide, which was demonstrated by XRD test (Fig. 12.5) and TPR test (Fig. 12.6). While excessively high temperature would lead to low yield, which was probably due to catalyst sintering at the high temperature to result in the destruction of nanaostructure. The suitable reduction temperature was in the range from 500 °C to 600 °C. Under the energy saving consideration, we chose 500 °C as the best reduction temperature. From the results of H_2–TPR measurement, one broad peak around 500 °C could be detected. It indicated an intense interaction between Ni component and Al_2O_3, which explained why this high loading catalyst

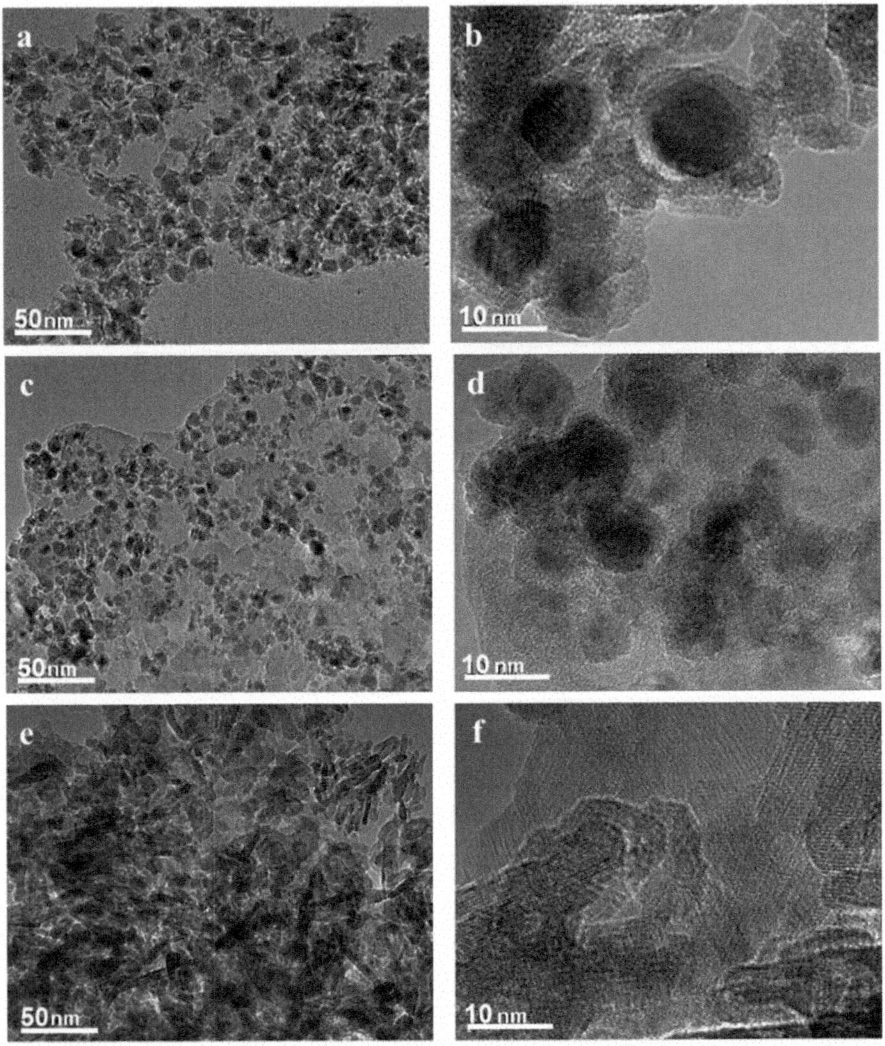

Fig. 12.4 TEM images of **a** Ni/Al_2O_3–HT, **c** Ni/Al_2O_3–CP and **e** Ni/Al_2O_3–IMP. HRTEM images of **b** Ni/Al_2O_3–HT, **d** Ni/Al_2O_3–CP and **f** Ni/Al_2O_3–IMP. Reprinted from Wiley (Ren et al., 2021)

still had such small Ni particle size. Neighboring alumina prevented Ni nanoparticle from further growing. Other solvents such as EtOH and MeOH instead of H_2O were examined, 56.7% and 74.3% yields of THFA were afforded, respectively (entries 11, 12). Among the other solvents such as DMSO, DMF, 1,4-dioxane, THF, acetone, DCM, ethyl acetate and n-hexane test, no desired products were detected (entries 13–20). The results clearly indicated that the solubility of $NaBH_4$ in solvent played a crucial role. Investigation of hydrogen sources, reaction temperature and time: The hydrogen sources were one of the major factors in the FA reduction process. In Table

Table 12.3 Physicochemical properties of Ni/Al_2O_3–HT

Entry	Ni/Al ratio	BET (m^2/g)	Pore diameter (nm)	Pore volume (cc/g)
1	0.25:1	156.7	2.1522	0.198
2	0.5:1	160.4	2.4417	0.209
3	1:1	157.8	2.8151	0.256
4	2:1	147.2	3.8938	0.288
5	3:1	145.6	8.7008	0.716

BET surface area determined from N_2 gas adsorption. Reprinted from Wiley (Ren et al., 2021).

12.5, the experiments were carried out with gaseous hydrogen and i-PrOH which give 68.9% and 64.5% yields of THFA, respectively (entries 2–3). The yield of afforded by the reaction using gaseous hydrogen was lower than that of $NaBH_4$, which was probably due to heterogeneous hydrogen source couldn't mix with the substrate completely. The reaction with hydrazine hydrate and triethyl silicane also gave very poor results (entries 4–5). Isopropyl alcohol and hydrazine hydrate required extra steps to produce hydrogen gas to further reduce the yield of THFA. As for worse performance of triethyl silicane, we speculated that polymerization product might be generated during reaction and covered the surface of catalyst to prevent further reaction.

The effects of other reaction parameters such as temperature and time were also investigated. As shown in Fig. 12.7, the reaction was conducted from 20 °C to 100 °C for 90 min. 60 oC was a mild and suitable temperature which could give 100% yield of THFA after 90 min reaction. Then, the time test was carried out at 60 °C, at the early stage of the reaction, the FA was converted quickly and reached a maximum value of THFA at 60 min. Thus, the optimal reaction temperature and time were 60 °C and 60 min, respectively. Compared with noble and other Ni-based catalysts for hydrogenation of FA in H donors, solvents, reaction temperature, time and yield of THFA, which were reported in recent years, Ni/Al_2O_3–HT also diplays excellent performance (Table 12.6).

Reaction conditions: 0.23 mmol substrate, 10 mg Ni/Al_2O_3–HT (Ni/Al = 2:1), 15 mg $NaBH_4$, 60 °C, 2 mL H_2O, 60 min. Reprinted from Wiley (Ren et al., 2021).

Recycling test of Ni/Al_2O_3–HT: As shown in Fig. 12.8, the stability test for Ni/Al_2O_3–HT was conducted at least 3 times which the yield of THFA slightly decreased. In previous reports, when Ni-based catalysts were used with H_2, after recycling several times, byproducts such as 2-methyltetrahydrofuran and tetrahydrofuran were always detected. But in this study, these by-products were not detected after recycled several times with $NaBH_4$ as hydrogen source. This may be due to the weaker destructive effect of $NaBH_4$ than that of H_2. ICP test showed that only 0.7221 ± 0.087 ppm Ni was detected, indicating that Ni didn't leach during the reaciton. As shown in Figs. 12.9 and 12.10, SEM and TEM images of catalysts indicated that no significant changes were observed before and after the reaction.

Scopes of substrates: With optimized reaction conditions in hand, the reduction of the substituted furan ring compounds was also explored, and the results were

Table 12.4 Optimization of Ni/Al ratio, reduction temperature and solvents

Entry	Ni/Al ratio	Catalyst loading (mg)	Reduction temperature (°C)	Solvent	THFA Yield (%)
1	0.25:1	10	500	H_2O	78.3
2	0.5:1	10	500	H_2O	89.1
3	1:1	10	500	H_2O	93.1
4	2:1	10	500	H_2O	100
5	3:1	10	500	H_2O	92.7
6	2:1	5	500	H_2O	92.9
7	2:1	10	300	H_2O	6.8
8	2:1	10	400	H_2O	7.0
9	2:1	10	600	H_2O	100
10	2:1	10	800	H_2O	98.2
11	2:1	10	500	MeOH	74.3
12	2:1	10	500	EtOH	56.7
13	2:1	10	500	1,4-dioxane	0
14	2:1	10	500	DMSO	0
15	2:1	10	500	DMF	0
16	2:1	10	500	Acetone	Trace
17	2:1	10	500	THF	0
18	2:1	10	500	CH_2Cl_2	0
19	2:1	10	500	AcOEt	0
20	2:1	10	500	n-hexane	0

Reaction conditions: 20 μL FA, catalyst Ni/Al_2O_3–HT, 15 mg $NaBH_4$, 100 °C, 2 mL solvent, 90 min. Reprinted from Wiley (Ren et al., 2021).

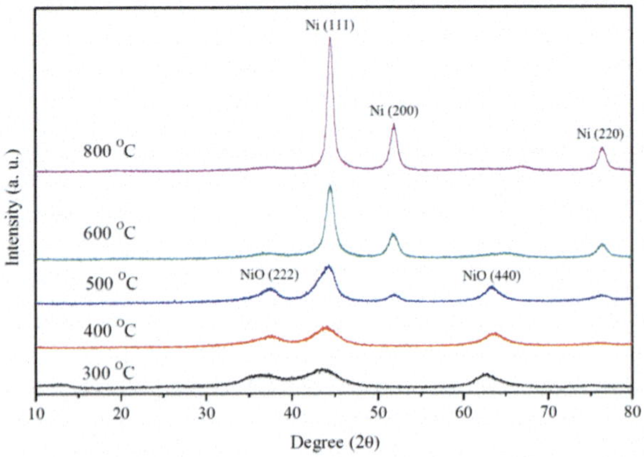

Fig. 12.5 XRD patterns of Ni/Al_2O_3–HT prepared at different reduction temperature. Reprinted from Wiley (Ren et al., 2021)

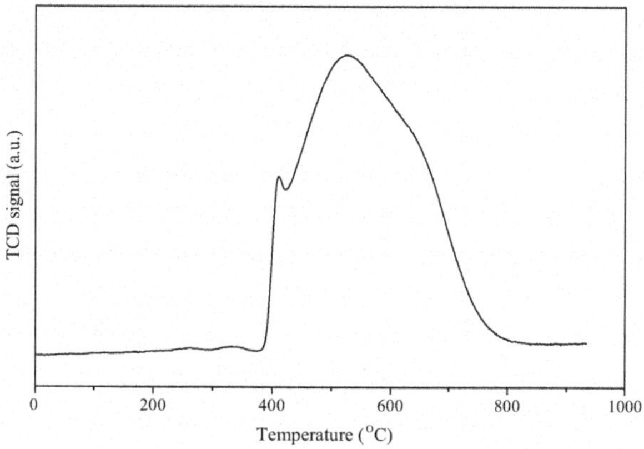

Fig. 12.6 H_2–TPR profile of Ni/Al_2O_3–HT (Ni:Al = 2:1). Reprinted from Wiley (Ren et al., 2021)

Table 12.5 Optimization of hydrogen sources

Entry	Hydrogen sources	Amount	THFA yield (%)
1	$NaBH_4$	15 mg	100
2	H_2	1.3 MPa	68.9
3	i-PrOH	30 μL	64.5
4	$N_2H_4{\cdot}H_2O$	30 μL	9.3
5	$HSiEt_3$	30 μL	4.2

Reaction conditions: 20 μL FA, 10 mg Ni/Al_2O_3–HT (Ni/Al = 2:1), 100 °C, 2 mL H_2O, 90 min. Reprinted from Wiley (Ren et al., 2021).

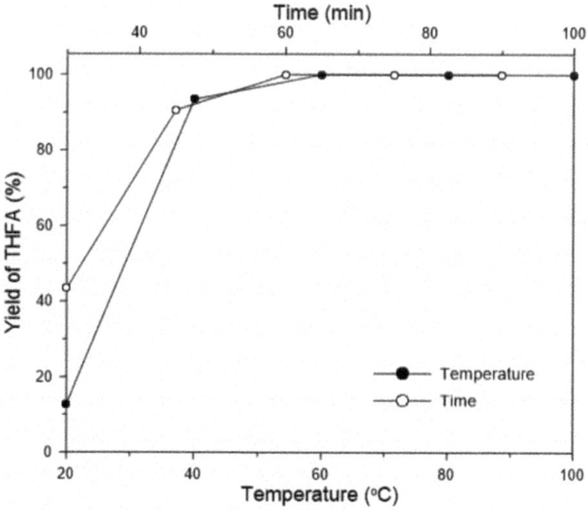

Fig. 12.7 Optimization of reaction temperature and time. (Reaction conditions: 20 μL furfuryl alcohol, 10 mg Ni/Al_2O_3–HT(Ni/Al = 2:1), 15 mg $NaBH_4$, 2 mL H_2O, for temperature investigation, time is 90 min; for time investigation, temperature is 60 °C.). Reprinted from Wiley (Ren et al., 2021)

Table 12.6 Scope of substrates

Entry	Substrate	Product	Yield (%)
1	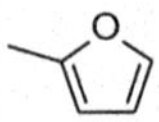	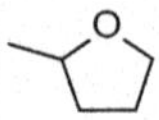	12.7
2	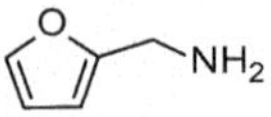	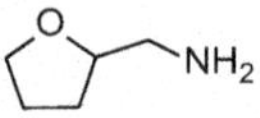	69.3
3	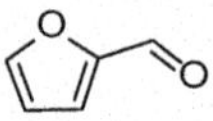	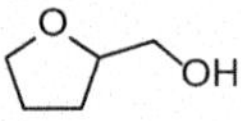	95.2
4	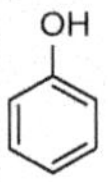	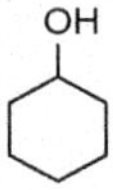	16.4
5	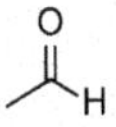	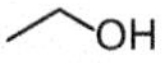	93.4
6	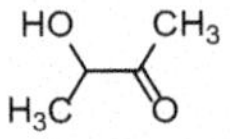	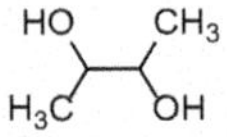	100
7	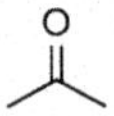	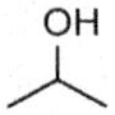	100
8	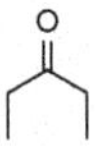		100

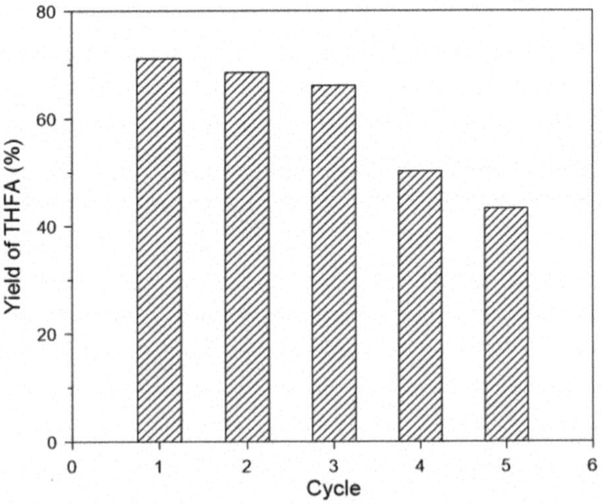

Fig. 12.8 Recycling test of Ni/Al_2O_3-HT. (Reaction conditions: 60 μL furfuryl alcohol, 10 mg Ni/Al_2O_3–HT(Ni/Al = 2:1), 15 mg $NaBH_4$, 2 mL H_2O, 60 °C, 90 min). Reprinted from Wiley (Ren et al., 2021)

Fig. 12.9 SEM images of Ni/Al_2O_3–HT catalyst. Left: before reaction; Right: after being recycled. Scale bars: 500 nm. Reprinted from Wiley (Ren et al., 2021)

summarized in Table 12.6. The reaction of 2-furfurylamine and furfural (entries 2–3), having aminomethyl group and aldehyde group in furan ring produced smoothly to give desired products tetrahydrofurfurylamine and THFA in 69.3% and 95.2% yield. Clearly, electron withdrawing groups decreased electron density of furan ring, which made it easier for hydride ion attacking furan ring. In contrast, 2-methylfuran gave desired products in very low yield (entry 1). We think that this may due to small water solubility and electron donating group. The reduction of benzene ring was investigated (entry 4), but the yield of 16.4% was less than satisfactory. Then the reduction of aldehyde group was also investigated. The substrates including acetaldehyde, acetoin, acetone and cyclohexanone, afforded desired products in very good yield (entries 5–8). Notably, the smaller the chemical shift of carbonyl carbon, the lower the yield. For example, acetaldehyde had the smallest carbonyl carbon chemical shift of this four carbonyl compounds (Hawkes et al., 1974), and the yield was the lowest of this four compounds. The result was in accordance with our above speculation, lower electron density was beneficial for hydride ion to attack substrates.

Proposed mechanism: According to present results and previous works (Dalla et al., 1999), a plausible mechanism is illustrated in Fig. 12.11. Initially, H^{+rds} in water attacked furfuryl alcohol molecule, and an electric intermediate was formed. At the same time, B-H bond cleavage took place in water, then the generated hydride ion absorbed on the surface of Ni particle to give the [Ni–H] species (Fountoulaki et al., 2014; Conte et al., 2009). After [Ni–H] species attacked the electric intermediate, Ni particle was regenerated. Through nucleophilic addition reaction, these two kinds of hydrogen atom were added into furan ring, and desired product THFA was formed.

We considered that catalyst support played a crucial role in reduction process. Metal ion of support reduced electron density of furan ring, which promoted interaction between furan ring and hydride ion. We also believed that hydride [Ni–H] possess much higher reaction activity than hydride ion, because this reaction didn't occur without catalyst. The presence of hydrogen ion seems to play a crucial role in this reaction, which explained why water as solvent gave the best yield.

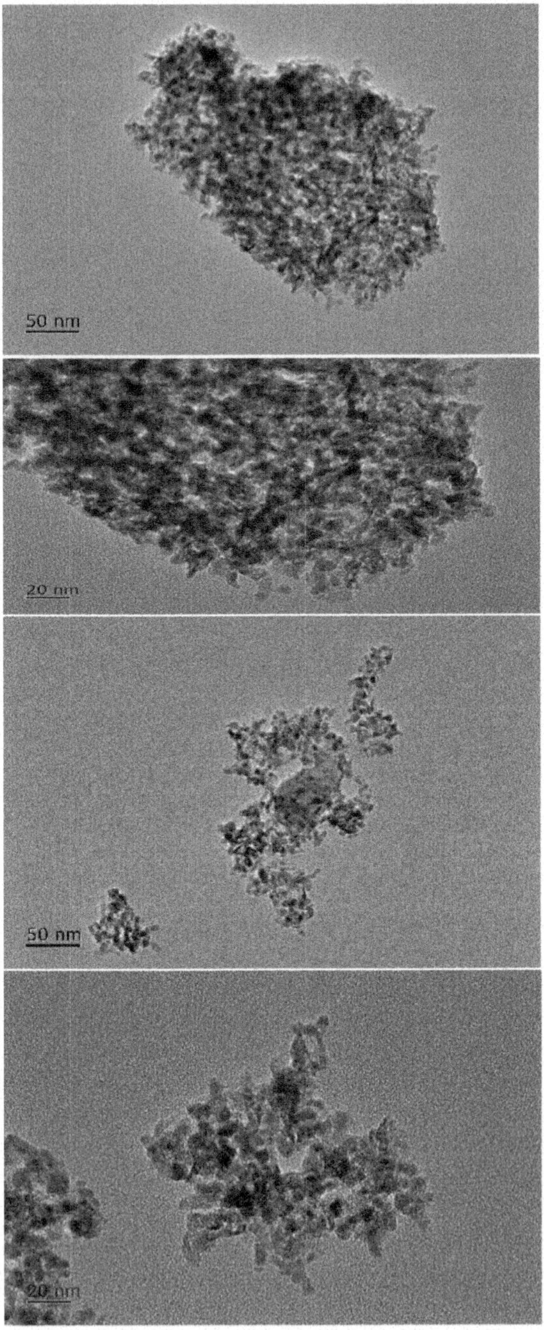

Fig. 12.10 TEM images of Ni/Al_2O_3–HT catalyst: up: before reaction; down: after being recycled. Scale bars: 50 nm and 20 nm. Reprinted from Wiley (Ren et al., 2021)

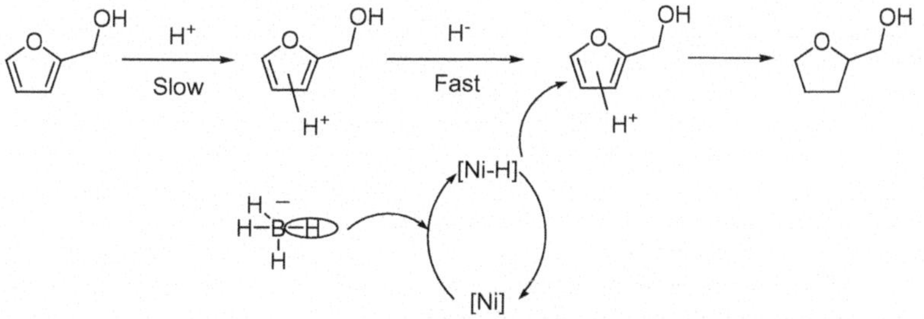

Fig. 12.11 Proposed reaction mechanism. Reprinted from Wiley (Ren et al., 2021)

12.3 Conclusions

In summary, aqueous-phase catalytic conversion of FA to THFA was investigated over a good dispersion Ni catalyst. Different from other preparing methods, catalyst prepared by Ni–Al hydrotalcite-like compounds exhibited high catalytic activity. With optimized conditions, the maximum THFA value of 100% was obtained in the presence of Ni/Al_2O_3–HT and 2 mL H_2O at 60 °C for 60 min. The method reported in this work showed good cycling performance and gave excellent yield at mild reaction conditions. $NaBH_4$ was first used for the reduction of FA, and Ni–Al hydrotalcite derived catalyst could be a promising candidate to replace noble metals to reduce FA under ambient conditions.

References

Biradar, N. S., Henhne, A. A., Birajdar, S. N., Swami, R., & Rode, C. V. (2014). Tailoring the product distribution with batch and continuous process options in catalytic hydrogenation of furfural. *Organic Process Research and Development, 18*, 1434–1442.

Cao, Y., Niu, L., Wen, X., Feng, W., Huo, L., & Bai, G. (2016). Novel layered double hydroxide/oxide-coated nickel-based core–shell nanocomposites for benzonitrile selective hydrogenation: An interesting water switch. *Journal of Catalysis, 339*, 9–13.

Cao, Y., Zhang, H., Dong, J., Ma, Y., Sun, H., Niu, L., Lan, X., Cao, & L., Bai, G. (2019). A stable nickel-based catalyst derived from layered double hydroxide for selective hydrogenation of benzonitrile. *Molecular Catalysis, 475*, 110452.

Chen, X., Sun, W., Xiao, N., Yan, Y., & Liu, S. (2007). Experimental study for liquid phase selective hydrogenation of furfuryl alcohol to tetrahydrofurfuryl alcohol on supported Ni catalysts. *Chemical Engineering Journal, 126*, 5–11.

Chen, Y., Yan, J., Ouyang, D., Qian, L., Han, L., & Chen, M. (2017). Heterogeneously catalyzed persulfate by CuMgFe layered double oxide for the degradation of phenol *Applied Catalysis A-General, 538*, 19–26.

Conte, M., Miyamura, H., Kobayashi, S., & Chechik, V. (2009). Spin trapping of Au-H intermediate in the alcohol oxidation by supported and unsupported gold catalysts. *Journal of the American Chemical Society, 131*, 7189–7196.

Corma, A., Iborra, S., & Velty, A. (2007). Chemical routes for the transformation of biomass into chemicals. *Chemical Reviews, 107*, 2411–2502.

Cui, X., Surkus, A. E., Junge, K., Topf, C., Radnik, J., Kreyenschulte, C., & Beller, M. (2016). Highly selective hydrogenation of arenes using nanostructured ruthenium catalysts modified with a carbon–nitrogen matrix. *Nature Communications, 7*, 11326.

Dalla, V., Catteau, J. P., & Pale, P. (1999). Mechanistic rationale for the $NaBH_4$ reduction of α-keto esters. *Tetrahedron Letters, 40*, 5193–5196.

Das, D., Roy, S., & Das, P. K. (2004). Efficient and simple $NaBH_4$ reduction of esters at cationic micellar surface. *Organic Letters, 6*, 4133–4136.

Dong, J., Zhu, T., Li, H., Sun, H., Wang, Y., Niu, L., Wen, X., & Bai, G. (2019). Biotemplate-Assisted synthesis of layered double oxides confining ultrafine Ni nanoparticles as a stable catalyst for phenol hydrogenation. *Industrial and Engineering Chemistry Research, 58*, 14688–14694.

Fan, G., Li, F., Evans, D. G., & Duan, X. (2014). ChemInform abstract: Catalytic applications of layered double hydroxides: Recent advances and perspectives. *Chemical Society Reviews, 43*, 7040–7066.

Fang, W., Xing, M., & Zhang, J. (2014). A new approach to prepare Ti^{3+} self-doped TiO_2 via $NaBH_4$ reduction and hydrochloric acid treatment. *Applied Catalysis B: Envitonment, 160–161*, 240–246.

Fountoulaki, S., Daikopoulou, V., Gkizis, P. L., Tamiolakis, I., Armatas, G. S., & Lykakis, I. N. (2014). Mechanistic studies of the reduction of nitroarenes by NaBH 4 or hydrosilanes catalyzed by supported gold nanoparticles. *ACS Catalysis, 4*, 3504–3511.

Furusawa, T., Shirasu, M., Sugiyama, K., Sato, T, Itoh, N., & Suzuki, N. (2016). Preparation of Ru/ZrO_2 catalysts by $NaBH_4$ reduction and their catalytic activity for NH_3 decomposition to produce H_2. *Industrial and Engineering Chemistry Research, 55*, 12742–12749.

Gao, L., Sun, J., Xu, W., & Xiao, G. (2017). Catalytic pyrolysis of natural algae over Mg-Al layered double oxides/ZSM-5 (MgAl-LDO/ZSM-5) for producing bio-oil with low nitrogen content. *Bioresource Technology, 225*, 293–298.

Geilen, F. M. A., Engendahl, B., Winterle, S., Liauw, M. A., Klankermayer, J., & Leitner, W. (2011). Selective homogeneous hydrogenation of biogenic carboxylic acids with $[Ru(TriPhos)H]^+$: a mechanistic study. *Journal of the American. Chemical Socirty, 123*, 6963–6966.

Ghaderi, S., & Setamdideh, D. (2014). $NaBH_4/NaNO_3/H_2O$: A convenient system for selective reduction of aldehydes VS. ketones to their corresponding alcohols. *Oriental Journal of Chemistry, 30*, 1913–1917.

Gong, L. -H., Cai, Y. -Y., Li, X.-H., Zhang, Y. -N., Su, J., & Chen. J. -S. (2014). Room-temperature transfer hydrogenation and fast separation of unsaturated compounds over heterogeneous catalysts in an aqueous solution of formic acid. *Green Chemistry, 16*, 3746–3751.

Gowda, A. S., Parkin, S., & Ladipo, F. T. (2012). Hydrogenation and hydrogenolysis of furfural and furfuryl alcohol catalyzed by ruthenium(II) bis(diimine) complexes. *Applied Organometallic Chemistry, 26*, 86–93.

Hawkes, G. E., Herwig, K., & Roberts, J. D. (1974). Nuclear magnetic resonance spectroscopy. Use of 13C spectra to establish configurations of oximes. *Journal of Organic Chemistry, 39*, 1017–1028.

Hronec, M., Fulajtarova, K., & Soták, T. (2014). Highly selective rearrangement of furfuryl alcohol to cyclopentanone. *Applied Catalysis B: Environmental, 154–155*, 294–300.

Hu, X., Westerhof, R. J. M., Wu, L., Dong, D., & Li, C. -Z. (2015). Upgrading biomass-derived furans via acid-catalysis/hydrogenation: The remarkable difference between water and methanol as the solvent. *Green Chemistry, 17*, 219–224.

Huber, G. W., Iborra, S., & Corma, A. (2006). Synthesis of transportation fuels from biomass: chemistry, catalysts, and engineering. *Chemical Reviews, 106*, 4044–4098.

Jiao, Z. -F., Guo, X. -N., Zhai, Z. -Y., Jin, G. -Q., Wang, X. -M., & Guo, X. -Y. (2014). The enhanced catalytic performance of Pd/SiC for the hydrogenation of furan derivatives at ambient temperature under visible light irradiation. *Catalysis Science and Technology, 4*, 2494–2498.

Ju, L., Wu, P., Lai, X., Yang, S., Gong, B., Chen, M., & Zhu, N. (2017). Synthesis and characterization of fullerene modified ZnAlTi-LDO in photo-degradation of Bisphenol A under simulated visible light irradiation. *Environmental Pollution, 228*, 234–244.

Karinen, R., Vilonen, K., & Niemela, M. (2011). Biorefining: heterogeneously catalyzed reactions of carbohydrates for the production of furfural and hydroxymethylfurfural. *ChemSusChem, 4*, 1002–1016.

Khan, F. -A., Vallat, A., Süss-Fink, G. (2011). Highly selective low-temperature hydrogenation of furfuryl alcohol to tetrahydrofurfuryl alcohol catalysed by hectorite-supported ruthenium nanoparticles. *Catalysis Communications, 12*, 1428–1431.

Khanapure, S. P., Saha, G., Sivendran, S., Powell, W. S., & Rokach, J. (2000). Intramolecular sulfur-assisted NaBH 4 reduction of esters synthesis of 5-oxo-ETE and 5-oxo-12-HETE. *Tetrahedron Letters, 41*, 5653–5657.

King, G. M., Iqbal, S., Miedziak, P. J., Brett, G. L., Kondrat, S. A., Yeo, B. R., Liu, X., Edwards, J. K., Morgan, D. J., Knight, D. K., & Hutchings, G. J. (2015). An investigation of the effect of the addition of tin to 5 %Pd/TiO_2 for the hydrogenation of furfuryl alcohol. *ChemCatChem, 7*, 2122–2129.

Koso, S., Furikado, I., Shimao, A., Miyazawa, T., Kunimori, K., & Tomishige, K. (2009). Chemoselective hydrogenolysis of tetrahydrofurfuryl alcohol to 1,5-pentanediol. *Chemical Communications*, 2035–2037.

Lee, J., Xu, Y., & Hube, G. W. (2013). High-throughput screening of monometallic catalysts for aqueous-phase hydrogenation of biomass-derived oxygenates. *Applied Catalysis B: Environmental, 140–141*, 98–107.

Lei, C., Pi, M., Kuang, P., Guo, Y., & Zhang, F. (2017). Organic dye removal from aqueous solutions by hierarchical calcined Ni–Fe layered double hydroxide: Isotherm, kinetic and mechanism studies. *Journal of Colloid and Interface Science, 496*, 158–166.

Lin, M., Yu, X., Yang, X., Li, K., Ge, M., & Li, J. (2017). Highly active and stable interface derived from Pt supported Ni/Fe layered double oxides for HCHO oxidation. *Catalysis Science and Technology, 7*, 1573–1580.

Lower, E. S. (1995). Speciality corrosion inhibitors. *Anti-Corrosion Methods and Material, 42*, 23–27.

Ma, H., Wang, H., Wu, T., & Na, C. (2016). Highly active layered double hydroxide-derived cobalt nano-catalysts for p-Nitrophenol reduction. *Applied Catalysis B: Environmental, 180*, 471–479.

Panagiotopoulou, P., & Vlachos, D. G. (2014). Liquid phase catalytic transfer hydrogenation of furfural over a Ru/C catalyst. *Applied Catalysis A-General, 480*, 17–24.

Pearson, R. G. (1963). Hard and soft acids and bases. *Journal of American Chemical Society, 85*, 3533–3539.

Peng, H. H., Chen, J., de Jiang, Y., Guo, X. L., Chen, H., & Zhang, Y. X. (2016). Merging of memory effect and anion intercalation: MnOx-decorated MgAl-LDO as high-performance nano-adsorbent for the removal of methyl orange. *Dalton Transactions, 45*, 10530–10538.

Ren, D., Wang, J., Jiang, X., Song, Z., & Huo, Z. (2021). A supported Ni catalyst produced from Ni–Al Hydrotalcite-like precursor for reduction of furfuryl alcohol to tetrahydrofurfuryl alcohol by $NaBH_4$ in water. *ChemistrySelect, 6*, 551–556.

Sang, S., Wang, Y., Zhu, W., & Xiao, G. (2016). Selective hydrogenation of furfuryl alcohol to tetrahydrofurfuryl alcohol over Ni/γ-Al_2O_3 catalysts. *Research on Chemical Intermediates, 43*, 1179–1195.

Sato, S., Igarashi, J., & Yamada, Y. (2013). Stable vapor-phase conversion of tetrahydrofurfuryl alcohol into 3,4-2H-dihydropyran. *Applied Catalysis A-General, 453*, 213–218.

Suzuki, K., Tanaka, N., Kageyama, K., & Takagi, H. (2011). Fabrication of well-dispersed barium titanate nanoparticles by the electrospray of a colloidal solution. *Journal of Materials Research, 24*, 1543–1552.

Tike, M. A., & Mahajani, V. V. (2007). Kinetics of liquid-phase hydrogenation of furfuryl alcohol to tetrahydrofurfuryl alcohol over a Ru/TiO_2 catalyst. *Industrial and Engineering Chemistry Research, 46*, 3275–3282.

Wang, Q., & O'Hare, D. (2012). Recent advances in the synthesis and application of layered double hydroxide (LDH) nanosheets. *Chemical Reviews, 112*, 4124–4155.

Wang, Y., Li, Q., Bo, L., Wang, X., Zhang, T., Li, S., Ren, P., & Wei, G. (2016). Synthesis and oil absorption of biomorphic MgAl layered double oxide/acrylic ester resin by suspension polymerization. *Chemical Engineering Journal, 284*, 989–994.

Wang, Y., Zhao, D., Rodríguez-Padrón, D., & Len, C. (20190. Recent advances in catalytic hydrogenation of furfural. *Catalysts, 9*, 796.

Wei, W. -L., Zhu, H. -Y., Zhao, C. -L., Huang, M. -Y., & Jiang, Y. -Y. (2004). Asymmetric hydrogenation of furfuryl alcohol catalyzed by a biopolymer–metal complex, silica-supported alginic acid–amino acid–Pt complex. *Reactive Functional Polymers, 59*, 33–39.

Yang, Y., Du, Z., Huang, Y., Lu, F., Wang, F., Gao, J., & Xu, J. (2013). Conversion of furfural into cyclopentanone over Ni–Cu bimetallic catalysts. *Green Chemistry, 15*, 1932.

Yuan, Q. , Ye, F., Xue, T., & Guan, Y. (2015). Room temperature hydrogenation of furfuryl alcohol by Pd/titanate nanotube. *Applied Catalysis A-General, 507*, 26–33.

Zhang, B., Zhu, Y., Ding, G., Zheng, H., & Li, Y. (2012). Selective conversion of furfuryl alcohol to 1,2-pentanediol over a Ru/MnOx catalyst in aqueous phase. *Green Chemistry, 14*, 3402.

Zhang, X., Tang, Y., Qiao, N., Li, Y., Qu, S., & Hao, Z. (2015). Comprehensive study of H_2S selective catalytic oxidation on combined oxides derived from Mg/Al-$V_{10}O_{28}$ layered double hydroxides. *Applied Catalysis B: Environmental, 176–177*, 130–138.

Zhang, H., Dong, J., Qiao, X., Qin, J., Sun, H., Wang, A., Niu, L., & Bai, G. (2019). In-situ generated highly dispersed nickel nanoclusters confined in MgAl mixed metal oxide platelets for benzoic acid hydrogenation. *Journal of Catalysis, 372*, 258–265.

Zhang, H., Guo, X., & Wang, X. (2020). Noble metal nanoclusters-decorated nife layered double hydroxide superstructure as nanoreactors for selective hydrogenation catalysis. *Nanoscale, 12*, 17780–17785.

Zhao, D., Su, T., Wang, Y., Varma, R. S., & Len, C. (2020). Recent advances in catalytic oxidation of 5-hydroxymethylfurfural. *Molecular Catalysis, 495*, 111133.

Zhou, Z., Tang, Y., Wang, L., Zhao, G., Zhou, Q., & Tang, C. (2004). Synthesis of optically active α,β-unsaturated triazolyl alcohols via asymmetric $NaBH_4$ reduction of the corresponding ketones. *Synthetic Communications, 34*, 1359–1365.

The manufacturer's authorised representative in the EU is Springer Nature Customer Service Centre GmbH, Europaplatz 3, 69115 Heidelberg, Germany. If you have any concerns regarding our products, please contact ProductSafety@springernature.com

Printed and bound by CPI Group (UK) Ltd, Croydon, CR0 4YY
01/10/2026
02225855-0001